Studies in Contemporary and Historical Archaeology 2

Henry's Mill: The Historical Archaeology of a Forest Community

Life around a timber mill in south-west Victoria, Australia, in the early twentieth century

Peter Davies

BAR International Series 1558
2006

Published in 2016 by
BAR Publishing, Oxford

BAR International Series 1558

Studies in Contemporary and Historical Archaeology 2
Henry's Mill: The Historical Archaeology of a Forest Community

ISBN 978 1 84171 988 7

BAR Publishing is the trading name of British Archaeological Reports (Oxford) Ltd. British Archaeological Reports was first incorporated in 1974 to publish the BAR Series, International and British. In 1992 Hadrian Books Ltd became part of the BAR group. This volume was originally published by Archaeopress in conjunction with British Archaeological Reports (Oxford) Ltd / Hadrian Books Ltd, the Series principal publisher, in 2006. This present volume is published by BAR Publishing, 2016.

Printed in England

BAR
PUBLISHING

BAR titles are available from:

BAR Publishing
122 Banbury Rd, Oxford, OX2 7BP, UK
EMAIL info@barpublishing.com
PHONE +44 (0)1865 310431
FAX +44 (0)1865 316916
www.barpublishing.com

Forward

Studies in Contemporary and Historical Archaeology

Studies in Contemporary and Historical Archaeology is a new series of edited and single-authored volumes intended to make available current work on the archaeology of the recent and contemporary past. The series brings together contributions from academic historical archaeologists, professional archaeologists and practitioners from cognate disciplines who are engaged with archaeological material and practices. The series will include work from traditions of historical and contemporary archaeology, and material culture studies, from Europe, North America, Australia and elsewhere around the world. It will promote innovative and creative approaches to later historical archaeology, showcasing this increasingly vibrant and global field, and celebrating its diversity, through extended and theoretically engaged case studies.

Proposals are invited from emerging and established scholars interested in publishing in or editing for the series. Further details are available from the series editors: Email Dan.Hicks@bristol.ac.uk or Joshua.Pollard@bristol.ac.uk

In this, the second volume in the series, Peter Davies presents a nuanced account of 19th and 20th century forest sawmill communities in southern Victoria, Australia. Weaving together archaeological and historical data, issues of community development, isolation, integration, and consumption practices are sensitively explored. Not only does the volume make a valuable contribution to the historical archaeology of rural Australia, but it provides an extended case study for others studying the history and archaeology of temporary work communities elsewhere in the emerging modern world.

Dan Hicks and Joshua Pollard (University of Bristol)
Series Editors

For William and James
and for Susan

Table of Contents

List of Figures

List of Tables

List of Plates

Abbreviations

AA – Australian Archives
BMA – British Medical Association
BT – The Birregurra Times
CH – The Colac Herald
CR – The Colac Reformer
FCV – Forests Commission Victoria
GA – The Geelong Advertiser
LCC – Land Conservation Council, Victoria
TW – The Timber Worker (journal of the Amalgamated Timber Workers' Union of Australia, Victorian Branch)
VPRS – Victorian Public Record Series, Public Record Office of Victoria.
WSA – The Winchelsea Shire Advertiser

Acknowledgements

This book originally began as a doctoral thesis in the Archaeology Program at La Trobe University, and could not have been prepared without the help of numerous people, whose support I take the opportunity to acknowledge here.

Early encouragement from Norm Houghton helped guide the first tentative steps of my work in the Otways, while friends and colleagues provided invaluable help with fieldwork for which I remain deeply grateful. In particular I would like to thank Åsa Ferrier, Sharon Lane, Jonathon Howell-Meurs, Tom Rymer and Lorenne Wilks, along with Lyndal Thorne, Tania Hardy-Smith, Susan Lawrence, Rudy Frank and Nick Porch. I am also indebted to Sarah Howell-Meurs for her analysis of faunal material from the site, and to Adrienne Ellis for her later work on the toy fragments from Henry's Mill. Thanks are also due to my supervisor, Tim Murray, for his continual encouragement and willingness to find my own way. Ming Wei and Rudy Frank helped with all things technical and logistical. Funding for fieldwork was provided by a La Trobe University Humanities Research Grant, and an Australian Postgraduate Award provided financial support.

Ian Sheuval and David Rourke of the Department of Sustainability and Environment helped facilitate my work in the Otway forests, and Paul Barker guided me through the labyrinth of Forests Commission archives. John Seebeck shared his family's history with me, while Jim Biddle, George and Elsie Chamberlain, Norm Gardner, Ric Henry, Joyce Harding and Jim Washington shared their memories. Friends and colleagues in the Australian Forest History Society offered encouragement and support, and the Birregurra District Historical Centre generously provided permission to reproduce images from their collections. Special thanks must also go to Dan Hicks and Joshua Pollard for their vision and energy in creating this new series of Studies in Contemporary and Historical Archaeology. Finally, to Susan, whose help and support have been far beyond words, thank you.

Chapter 1

Introduction

One day in September, 1919, a journalist visited Henry's No.1 Mill and observed that:

> ...the scenery en route is equal to anything in the State. The tramline wends its way along the valleys which are veritable fairylands of ferns, while the mountain streams tinkle musically over their pebbly beds...maidenhair fern, delicate and drooping, edge the river bank and trail coolly in the limpid waters of the purling stream. Miles of this exquisite scenery are encountered before the traveller reaches the site of the mill which is a scene of striking activity...The screaming of the huge saws as they relentlessly tear to pieces the mighty trunks...echoes and re-echoes down the dim green aisles of the forest, and ever and anon a thunderous crash reverberates through the valley as another mighty monarch of the bush, fated to feed the rapacious screaming monster...crashes to the ground through the agency of man, to be shortly re-incarnated, in unlovely form, in the shape of an unimposing weatherboard cottage in equally sordid surroundings (BT 30 Sept. 1919).

The writer, an outsider to the mill community, contrasts in florid terms the beauty of the natural world with the brutality of human industry. Taking pleasure in 'the forest fairyland', he sets its quiet grace against the noisy power of the sawmill in its midst. But what of the insiders, the men who chopped, carted and milled the great trees? Many brought their families with them, creating communities deep in the forest. How did they respond to the natural environment, to the world beyond, and to each other? Workers built rough wooden huts in clearings carved from the forest. What other amenities did they need and create? The industry was labour-intensive, dirty, and dangerous. How did people cope when injury or illness struck? Timber tramlines provided a lifeline connecting mills to the outside world. They brought in people, goods and mail, and carried timber away to market. Where did people come from, and for how long did they stay? This study examines the social and material lives of these men, women and children. Focusing on Henry's No.1 Mill, it integrates archaeological, oral and documentary evidence to illuminate community life at a typical forest settlement in early twentieth-century Australia.

Timber-getting was vital to the success of the mining, railway, construction and shipping industries. The proliferation of bush sawmills and forest camps provided employment for thousands of men, created settlements, and encouraged the development of transport infrastructure in remote and difficult areas. Several thousand forest mills were established in Victoria between the 1850s and World War II, indicating the scale of the industry in this period. Research on the history of sawmills and tramways, however, has been dominated by industrial and technological concerns (e.g. Evans 1994; Houghton 1992; McCarthy, M. J. 1993). The social history of mill settlements has been largely neglected. Forest-based mills and camps were mostly short-term, ephemeral in nature, generating few written records of their operation. The archaeological record thus represents a critical resource for expanding knowledge of people in a major rural industry, but who were otherwise of little note. Physical evidence helps to reveal a forest-based way of life and industry which had all but disappeared by the 1940s. As the twentieth century has been largely ignored by archaeologists in Australia, interpretation of material culture from this period is a significant focus of attention. This study thus offers a substantive archaeological investigation of rural Australian life.

Wider issues of community development, isolation and integration, and consumption practices in remote areas are also addressed here. A focus on the community highlights not only the responses of mill residents to the natural world, but the social tensions, political allegiances and personal rewards of daily life at a bush mill. The constant negotiation of geographical isolation by mill residents with their integration into wider aspects of contemporary society is a vital theme of this study. Material evidence indicates the regular acquisition of mass-produced consumer goods by industrial wage earners. It also emphasizes, however, the importance of reuse and recycling as local expressions of improvisation and thrift, and personal responses to illness and injury.

Henry's No.1 Mill was established in 1904, deep in the watershed of the West Barwon River in south-west Victoria (Figures 1 and 2; Plate 1). It was connected by a timber tramline to the railhead at Forrest, ten kilometres to the north, but the mill's isolation meant that a permanent population of around 100 people lived on site. The mill settlement featured rough timber huts for single men and modest wooden houses for married men and their families, along with a boarding house, billiard room and stables. Children attended the mill school, which opened in 1909,

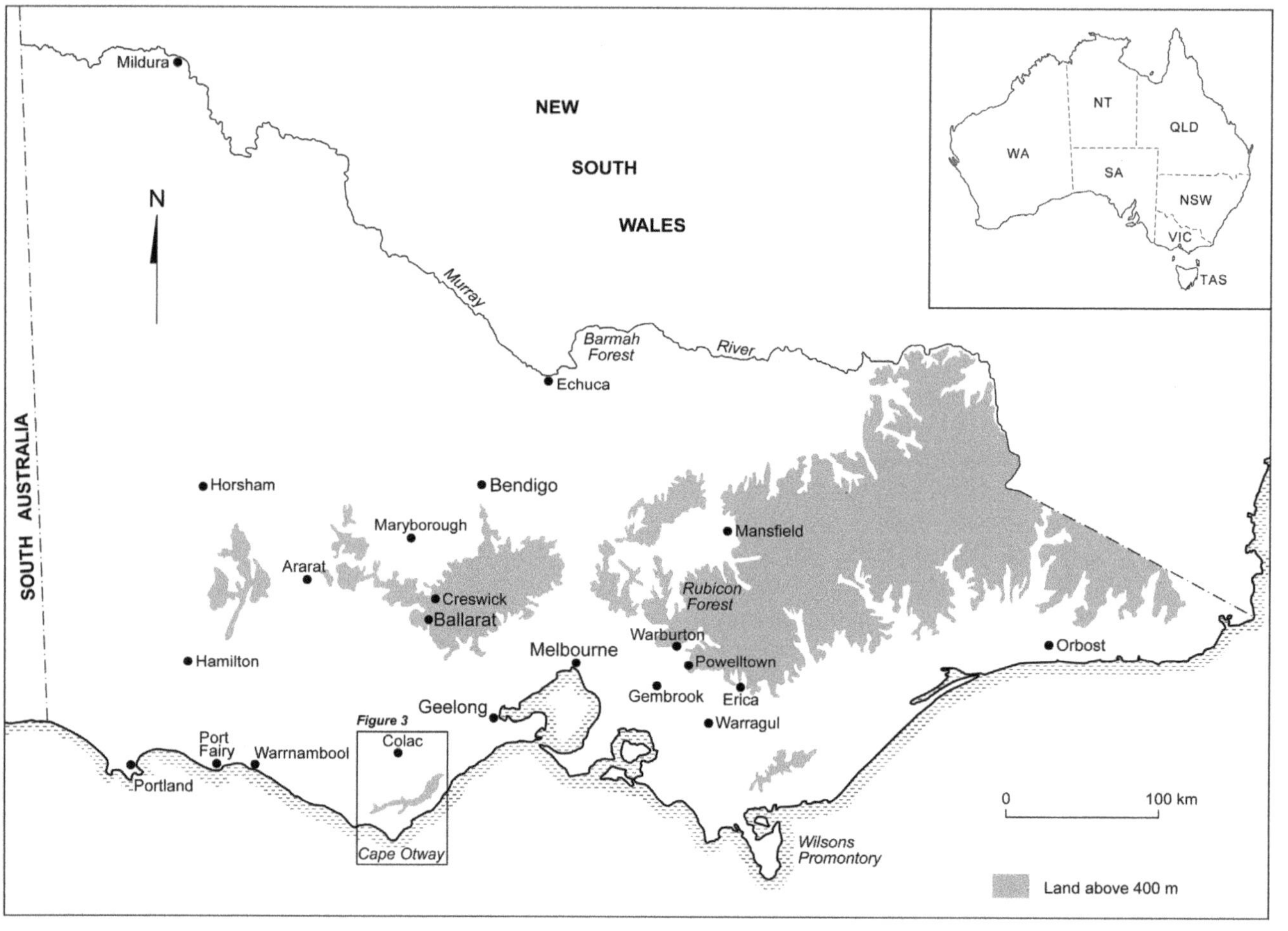

Figure 1. Map of Victoria, Australia showing location of Figure 3.

Plate 1. Contemporary view of Henry's Mill (photo: author).

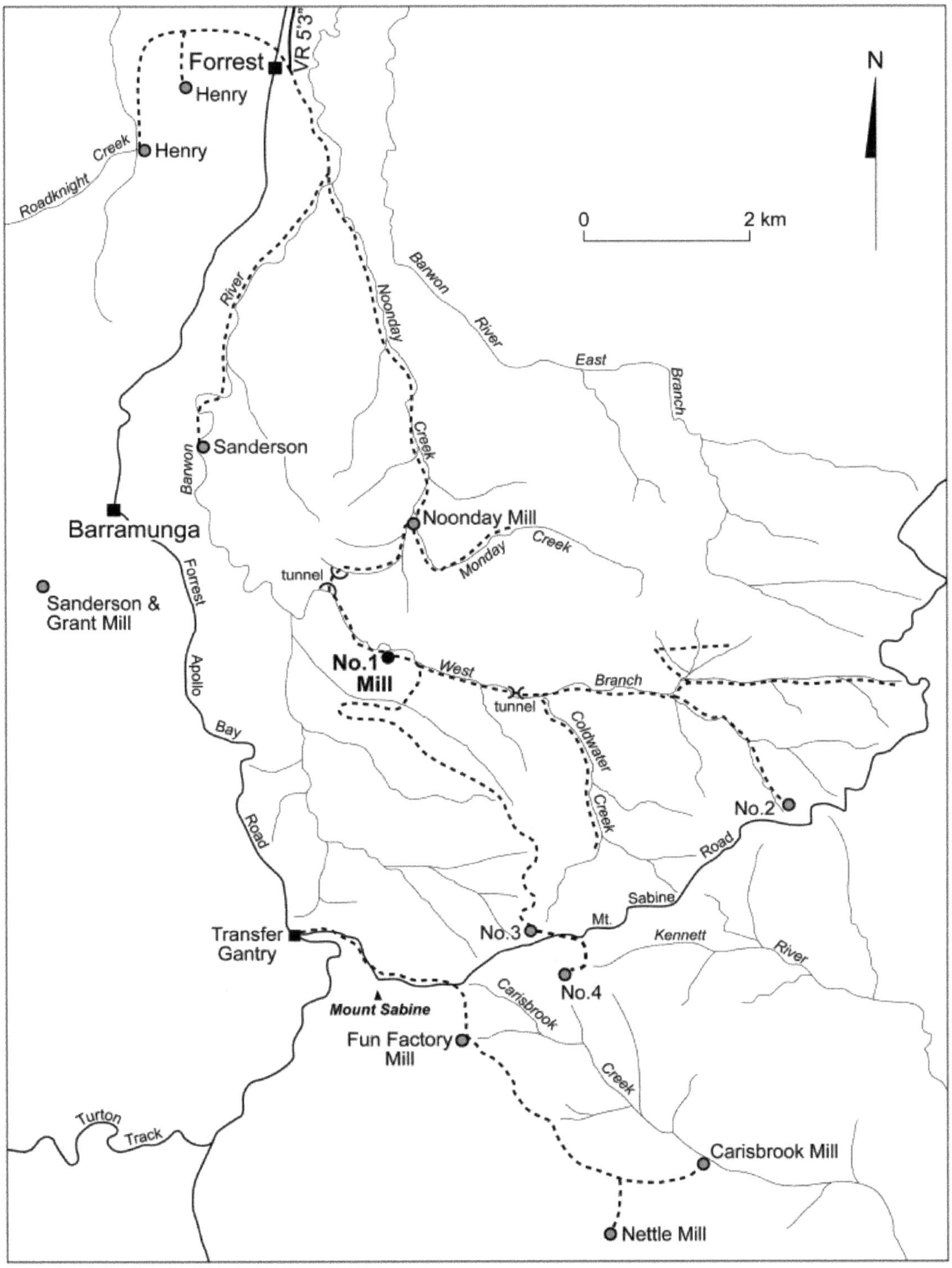

Figure 2. W.R. Henry & Son's tramlines (adapted from Houghton 1975).

while a post office and store provided further services. In later years the firm established several other mills higher up the valley as well. In 1927 the main mill shed was destroyed by fire, resulting in the gradual abandonment of the site.

Sustained over-cutting of forests during the nineteenth century left Victorian sawmillers with a much depleted resource, forcing them into increasingly remote areas where operating costs were higher. Railway expansion into the Otways from the late nineteenth century permitted the growth of a substantial timber industry in the region (Figure 3) Henry's No.1 Mill was thus characteristic of its era in being a forest-based sawmill located in the midst of an important timber resource. It operated at the high point of this mode of production, on the cusp of the transition from animal and steam power to diesel and electricity. Henry's was one of several mills in the area at which some portion of the crew lived permanently on site, featuring accommodation for workers and a boarding house to provide meals. Its mixture of single men and families was also typical of the industry. The

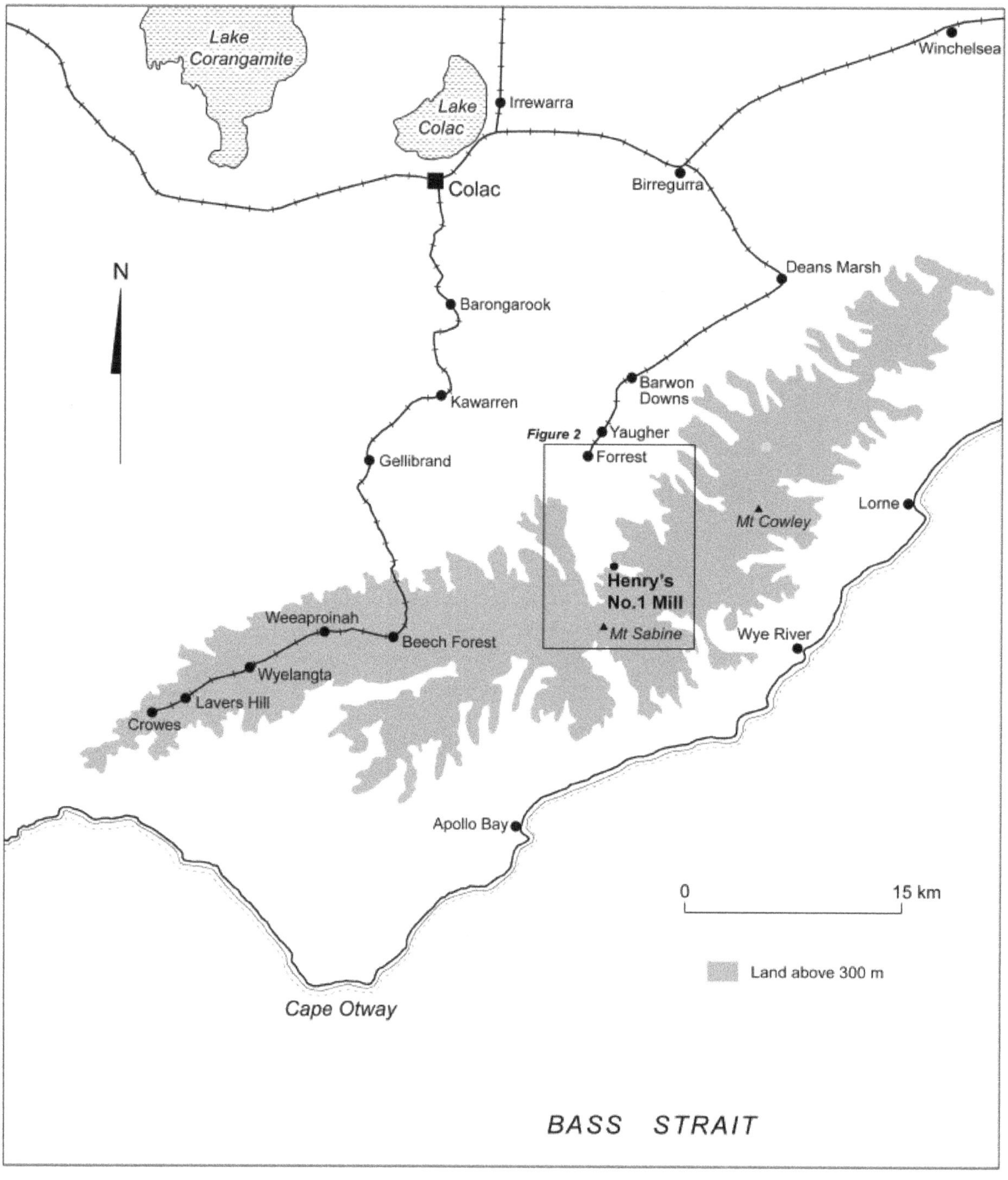

Figure 3. The Otway Railways

mill was unusual, however, in employing a large crew, producing a high timber output, and operating for a fairly lengthy period (about 24 years). The proprietors adapted available technology in innovative ways for the period, utilising locomotives, tram tunnels and steam winches. It was also one of the very few sawmills in the Otways to support its own school, and the only one known to have a post office.

The length of operation and number of people residing at the No.1 Mill meant there was a greater likelihood of documents and archaeological deposits being created and surviving to the present. The location of the site in a regional water catchment system precluded later disturbance by logging operations, thus maintaining a high level of integrity in its archaeological remains. The site was also, importantly, on the edge of logistical accessibility for field-based investigation. Its relatively recent demise in the late 1920s means that several informants with memories of the place have been available for interview. In addition, pioneering survey work conducted by Norm Houghton on the tramways and sawmills of the area have situated the No.1 Mill as a prominent place within a regional historical context (Houghton 1975, 1997). Building on Houghton's technical focus, this study represents a significant extension of his work.

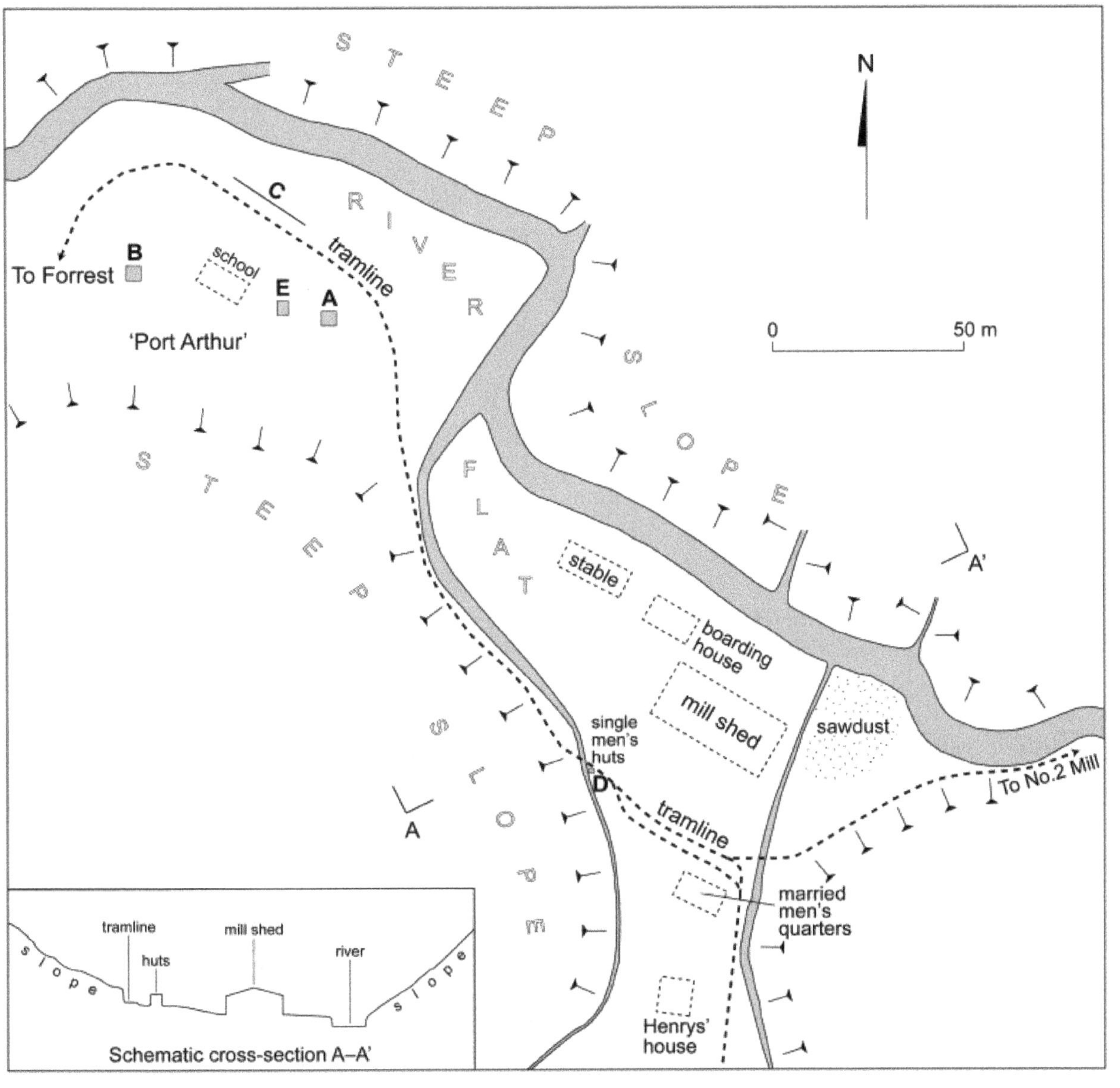

Figure 4. Plan of Henry's No. 1 Mill

Archaeological fieldwork at Henry's was conducted in two stages, in April-May, 1997, and January-February, 1998. The first involved intensive site survey to provide an accurate plan of the mill and its geographical and cultural boundaries, and to indicate areas of archaeological potential (Figure 4). Survey revealed substantial surface scatters representing occupation in the north-west of the site, but almost no domestic material in the area inhabited by unmarried workers. The second stage of fieldwork thus involved excavation of residential remains in the former area. This process aimed to recover material from household units relating to specific domains of inquiry. These included exchange networks linking the mill to the outside world; patterns of consumption and discard; the nature of domestic architecture and furnishings; and responses to the physical and social environment. Excavation concentrated on three house sites (House A, B and E), along with a small test pit in the area of the single men's huts (Sounding D), and an intensive surface collection below the tramline embankment along the river flat (Dump C).

Themes addressed in the Study

Five principal themes are addressed in the study, and are discussed below: *Community Archaeology, Isolation and Integration, Forest History, the Archaeology of the Twentieth Century* and *Archaeology and History*. In all cases, the approach adopted here owes less to a rigid program of hypothesis formulation and testing (e.g. South 1977; Lewis, K. E. 1984), than to the quieter, more intimate questioning of scholars like Henry Glassie (1982). Although Glassie's account of folklife at Ballymenone has been rightly criticised for its simple structuralist elements and the 'total immersion' technique of community observation, it nevertheless offers a compassionate and inspiring synthesis of culture, history, people, time and space. Glassie's study begins on 'the inside, where people are articulate and powerful and in control', allowing hypotheses to emerge naturally out of experience (Glassie 1982: 86). In a similar way, the themes pursued in the present study have developed significantly during the course of investigation. The accumulation and analysis of artefacts, anecdotes and other material has provided a rich source of evidence from which to draw and shape a portrait of people and their lives at the No.1 Mill.

Community Archaeology

This study uses the notion of 'community' to frame analysis of Henry's Mill. The nature of community has engaged thinkers from ancient times to the present. Plato, Aristotle, Hobbes, Marx and many others have explored how communities are created and maintained, and how individual needs are reconciled with the common good. The community concept in historical archaeology owes much to its use and development in sociology since the nineteenth century. Ferdinand Tönnies distinguished between community (*gemeinschaft*), which fulfilled the need for private, emotional relationships between people, and society (*gesellschaft*), characterised by colder, more impersonal connections to the marketplace (Tönnies 1963 [1887]: 33-102). Meanings of the term subsequently proliferated, with George Hillery identifying little common ground among the 94 definitions he catalogued (Hillery 1955). Human ecologists stressed space, habitat or settlement in their use of the term, while others focussed on collective psychological needs (see Poplin 1972: 65-107).

Formulations of community have also highlighted the importance of social interaction. Elias, for example, reinforces the residential aspects linking people to space, defining community as 'a group of households situated in the same locality and linked to each other by functional interdependencies' (Elias 1974: xix). Arensberg incorporates a temporal dimension as well, regarding the community as 'a structured social field of...relationships unfolding through time' (Arensberg 1961: 250). Warren refers to the social units and systems affording people opportunities to meet their everyday needs, while Poplin accepts the elements of geography, social interaction and common ties as central to notions of community (Poplin 1972: 9-27; Warren 1972: 9). Cohen treats community as a shared system of values, norms and moral codes which provide a sense of identity and boundedness to its members (Cohen 1985: 12).

More recently, Anderson has proposed the 'imagined community', in reference to the modern nation-state, and the 'image of communion' held by citizens who will never know each other (Anderson 1991:6). In a similar vein, Roseberry has defined community as a 'political association' formed through processes of political and cultural creation and imagination (Roseberry 1989: 14). Keller, however, defends the importance of personal attachment and belonging to a local place as the foundation of community. She highlights the value of both public and private spaces, with rules and norms to maintain order and civility, and the sharing of goals, traditions and values (Keller 2003: 293). Carroll, on the other hand, relates communities to consumption. She argues that the community provides a context for the exchange of commodities, and thus linkages between global networks and local places (Carroll, L. 1999)

In this study, 'community' refers broadly to a group of people interacting daily within a specific locale. A significant element of such community is personal, face-to-face interaction on a regular basis, facilitated by social institutions (church, school), workplace, residential proximity and kinship. The 'settlement' provides the physical spaces and built structures within which people interact and community is formed. The spatial organisation of a settlement reflects and structures the kinds of interaction possible between individuals (Lawrence 2000: 13-15). The community also shares a common subsistence and economic base, and participation in related exchange networks. Importantly, community involves an identification with place and a sense of belonging, a psychological dimension which can extend to shared cultural values and world-view. It involves a willingness to participate in communal life, and the security and fulfilment flowing from it. The community can also be characterised, however, by conflict, rivalry and hostility, as much as by cohesion, amity and unity of purpose.

One of the problems facing all community studies is their representativeness, the extent to which a particular people and place can be taken as characterising a society more generally. Conrad Arensberg asks

> How far can we use a known community of persons as a portrait, an epitome, a witness of the life, nature, fate, character, and spirit of the unknown others of the people of their age and civilization? (Arensberg 1961: 242)

Archaeologists in the United States have endeavoured to overcome this problem in various ways, integrating a tight focus on site-based material residues with broader historical themes and cultural patterns. Deagan's study of St. Augustine provides both a case history of an eighteenth-century Spanish American settlement, as well as an entry point for investigating Spain's colonisation of the New World (Deagan 1983). Geismar's archaeological analysis of a nineteenth-century black community in New Jersey offers insight into changing processes of social organisation and disintegration (Geismar 1982), while Lewis situates the town of Camden within broader processes of colonial expansion on the South Carolina frontier (Lewis, K. E. 1984). James Delle analysed the spatial design of early-nineteenth-century coffee plantations in Jamaica to understand broader changes in the world economy and their impact on local regions (Delle 1998). Investigation of the Boott Mills complex in Lowell, Massachusetts, revealed how the nineteenth-century built environment could reinforce ideologies of industrial capitalism and corporate paternalism (Beaudry 1989; Mrozowski, Ziesing and Beaudry 1996). Adams' analysis of Silcott, an early twentieth-century farming settlement in south-eastern Washington, examines the economic and social networks linking the people of the community to the products of the nation (Adams 1977, 1991). Cusick, in his important review of the community study approach in historical archaeology, argues that the strength of community analysis lies not only in its identification of social meaning within a specific locale, but in its interdisciplinary framework and its role in constructing a more complete ethnography of the past (Cusick 1995).

Archaeologists have also recently explored the notion of community at industrial work camps in the western United States (Van Bueren 2002). As the global economy expanded between the eighteenth and twentieth centuries, a variety of work communities developed to provide living quarters for workers extracting raw materials and building transport infrastructure. These industries included oil, coal and timber, along with dams, aqueducts and railways. Van Bueren is cautious, however, about calling these temporary work settlements 'communities', because they lacked the more diversified and sustainable economic foundations of agriculture, shipping or industry. Instead, such peripheral work settlements were defined by a narrow economic focus, relative geographic isolation, impermanence, and dependence on the global economy (Van Bueren 2002: 2). Archaeological investigation of work camps highlights the importance of studying peripheral settlements within the context of global developments, and the emerging relationships between work, capital and the environment (Hardesty 2002: 95-6).

Archaeological studies of Australian communities have also sought to link the particular to the general. Birmingham (1992) situated the remains of Aboriginal mission cottages at Wybalenna, on Flinders Island in Bass Strait, within the broader context of indigenous Tasmanians' interactions with European colonisation, ideology and consumerism. In the process of unearthing an ephemeral nineteenth-century mining community at Dolly's Creek, Lawrence identified such themes as family responsibility and domestic respectability, using these to challenge conventional historical notions of masculinity and Australian identity (Lawrence 1995, 2000). Investigations in the Rocks in Sydney were framed within a series of questions which related the neighbourhood to the city and beyond. These questions dealt with the impact of the Industrial Revolution on a pre-industrial town, changes in women's lives and experiences, standards of living and the role of government, while evidence was also used to contest the historical reputation of the Rocks as disease-ridden slum (Karskens 1999: 20; Karskens & Thorpe 1992). In a similar fashion, reanalysis of the 'Little Lon' precinct in Melbourne has offered a way of re-imagining a notional 'slum' neighbourhood as a vanished but once-vibrant working-class urban community (Mayne & Lawrence 1998; Murray and Mayne 2001, 2003).

This study faces a similar challenge, to situate Henry's Mill within a broader archaeological and historical context. The community approach applied here positions people and their material lives in social and geographical space. Synthesizing archaeological and documentary evidence, the approach helps control such variables as physical setting and commodity flow. As such, it allows for limited generalization on the basis of a specific case study (Cusick 1995: 60). The No.1 Mill typified forest-based work and life in early twentieth-century Victoria. Careful examination of archaeological and other evidence to document material and social life at the site can widen understanding of rural experience in this period. In their consumption practices, use of space and environmental responses, mill residents faced issues of resource utilisation and social interaction common to most other rural communities. I argue that the values and allegiances expressed by workers and residents at the mill reflect in microcosm many of the values common to the broader community of which they were a remote yet representative element.

Isolation and Integration

The archaeology of isolated settlements in the late nineteenth and early twentieth century is another significant area of inquiry relating to the present study. Conventional archaeological focus on the 'site' and its constituent living spaces is consistent with the community scale of analysis and comparison of assemblages at the household level. The community concept is thus well suited to analysis within historical archaeology, providing a viable framework to position people and their material world both socially and geographically (Cusick 1995: 60). It bridges the household scale with wider material and economic networks. Investigations at Dolly's Creek (Victoria), Silcott (Washington State) and the Moser farmstead (Arkansas) in particular have been influential in the present study, as they successfully combine archaeological, historical and ethnographic/oral sources of evidence to illuminate community life of non-elite working people. A significant concern in each case is to describe the material life of such people, and the relationships between community and broader spheres of economic and cultural interaction.

Silcott was a small farming community in south-eastern Washington which flourished between 1880 and 1930. W. H. Adams has examined the hierarchy of economic and social networks operating at the site in the early twentieth century (Adams 1977, 1991; see also Gaw 1975; Riordan & Adams 1985). By analysing the manufacturing origins of more than 1000 items, he identified co-operative labour exchange between neighbours, purchases in local and regional stores, connections to national manufacturing centres, and the role of mail order houses, as interwoven exchange processes linking the small farming community at Silcott to the rest of the United States and the world beyond. Although Adams blurs the distinction between individual acts of acquisition and the manufacturing origins of most products, his model highlights the connections between families and the community, and between communities and the nation. The study is limited, however, by its failure to explore the meaning of these exchanges and interactions within the lives of local inhabitants.

This concern with symbol and cognition, however, is carefully articulated by Stewart-Abernathy in his analysis of the Moser farmstead in north-west Arkansas (Stewart-Abernathy 1992; see also Purser 1992, 1999). Occupied from the 1880s to the 1920s, excavations at the site yielded a range of goods manufactured within a context of industrial mass-production. Quantities of canning jars, agricultural tools and ceramic tablewares, however, were at odds with the area's historical reputation as a remote, self-sufficient backwater. Investigators were forced to ask how the inhabitants negotiated the contradiction between agrarian and industrial world views, and the values each embodied. They observed that the selection, acquisition and use of goods frame people's lives with meaning. The choices of the Moser families indicated an allegiance both to the local community and the world beyond, but on terms which kept the community strong.

Lawrence identified a similar theme in her study of 'a poor man's diggings' on the Moorabool goldfields near Ballarat in the later nineteenth century (Lawrence 2000). The community of subsistence miners at Dolly's Creek followed traditional forms of organisation. Nevertheless, the industrial and commercial revolutions made mass-produced, inexpensive goods accessible even to such working people on the margins of settler societies. Families in the community maintained a dignified self-sufficiency largely outside the structure of capitalist wage labour, enlivening their rough homes with small touches of material comfort. The constant adaptation of long-standing customs with modern commodities reinforced their sense of identity, security and belonging.

However, Henry's Mill was both isolated from and integrated with the world beyond the forest. The interplay of these themes is central to the present study. Geoffrey Blainey has famously addressed the wider issue of distance and isolation in Australia in *The Tyranny of Distance* (1966). He argued that Australia's historical isolation from the rest of the world, and distances within and around the continent itself, have always limited and constrained the movement of people, ideas and materials. How these distances were gradually overcome, or 'tamed' during the course of the nineteenth century by steamships and railways, and later by aircraft and automobiles, is a central theme of the book. Blainey suggested that distance, and the isolation which resulted, is a crucial factor in explaining the development of the Australian economy, society and national character.

John Hirst challenged this view, arguing instead that people, goods and information have been highly mobile from the beginning of European settlement in Australia (Hirst 1975). Distances are relative, not absolute. The movement of commodities in a society is not related primarily to distance, but to the nature of goods produced, markets for them, and the means and costs of transport. The mobility of population depends on social attitudes and expectations, as well as on personal means and the availability of transport. People, materials and ideas, Hirst argues, interacted effectively in spite of the distances involved.

This historical tension, between distance and the isolation it engendered, and mobility and the integration it permitted, forms an important thematic framework for the study of sawmill settlements. Henry's No.1 Mill was not a frontier community; it was a rural industrial camp. Although isolated and somewhat self-contained, it was not entirely cut off from the outside world. A three-hour walk down the tramline to Forrest overcame the distance separating residents from a railway terminus and thence the world beyond. At the same time, the tramline offered the only practical way out, and it could not be undertaken at will. The constant negotiation of these two pressures, of isolation and integration, forms the underlying conceptual dynamic of the present study. Archaeological and other evidence will be used here to argue that workers and residents at Henry's responded to the force of isolation with a range of strategies. These included the adaptation of local materials to building and domestic needs, and the appropriation of commercial products to meet health and medical problems. In the end, most people solved the problem of geographical and social remoteness simply by leaving the mill in search of better opportunities elsewhere. Their integration with wider socio-economic structures, however, is clearly manifest in the wide range of manufactured goods brought to the settlement, as expressions of consumption by industrial workers.

Forest History

Forest history has emerged as a significant field of research in Australia in recent years, bringing together a heterogeneous mix of historians, geographers, archaeologists, foresters and others (e.g. Carver 2005; Dargavel and Libbis 1999; Dargavel, Gaughwin and Libbis 2002). The divergent approaches, materials and views they express, however, are broadly united by a concern to advance historical understanding of human interactions with forest and woodland environments. The discussion below reviews those aspects of forest history which relate specifically to sawmill communities: cultural heritage and the regional forest assessment process; light railway and timber tramway studies; and corporate history and the historical transformation of the timber economy in Australia.

Forest history in Australia operates at the intersection of natural and cultural heritage, where history and ecology converge. Cultural heritage has nevertheless only recently come to be widely accepted as an important dimension of forest value, especially in response to the Regional Forest Assessment process during the 1990s, along with environmental impact assessments (Lennon 1998). Cultural heritage studies have therefore increasingly adopted a regional perspective in recent years, to document and assess the places and relics of human behaviour in natural environments. Sites embodying cultural values in forests include sawmills and tramways, mines, dams, fire towers, lookouts and farms. These places provide cultural historians with evidence of past human activity, and ecologists with an understanding of environmental processes and change. The distinction between natural and cultural heritage is becoming increasingly blurred, with the recognition that today's forests, even those with old growth and wilderness values, are landscapes which incorporate Aboriginal occupation, historical timber-getting, grazing, mining and other activities (Griffiths 1996: 255-77; see also Schama 1995: 7-14).

The cultural heritage of forest areas has been documented for several regions of Victoria, including the Central Highlands (Evans 1993), the South-West (Story & Davies 1995), Gippsland (Evans 1998), and the North-East (Evans 1997b). In Tasmania, Kostoglou has identified many hundreds of forest activity sites in a number of forest management zones (e.g. Kostoglou 1992, 1994). The majority of such studies provide an inventory of historic forest industry sites, and assess the cultural heritage significance of each at local, regional, state or national levels. Along with outline histories, remnant cultural material and threats to the fabric of historic places are identified, and management objectives recommended. Some forest heritage studies aim to document as many sites as possible, providing a broad coverage of all the known forest activities in the study area. Others concentrate on a smaller sample of sites, assessing their significance against established criteria. The net result of such assessment processes has been to document and contextualise a large number of historic sites relating to sawmilling, timber transport and other forest industries. This 'dots-on-the-map' regional inventory approach provides an important tool for the management of forest heritage, and a starting point to extend our understanding of past human uses of forest and woodland environments. By canvassing wide areas, however, regional studies have been unable to develop detailed individual histories. This study helps to correct the imbalance.

Light railway studies share with cultural heritage assessment not only a similar suite of historical resources, but also a focus on forest environments, and a concern to document technologies and patterns of life no longer practised. In Australia, light railways and timber tramways have historically formed vital transport and communication links between a range of industrial activities and government railway lines, most prominently from the 1880s to the 1950s. Sawmills, construction projects, mines and canefields often utilised light railways and tramways to move people and materials from source to site, frequently pioneering the development of transport infrastructure in remote areas. In Victoria, numerous studies have considered the operation of timber tramway networks, built and maintained by sawmillers to permit transport of logs and timber from isolated forest mills to railheads. Districts covered include the east and west Otways (Houghton 1975, 1992), Upper Yarra (McCarthy 2001; Stamford et al. 1984), Erica (McCarthy, M. J. 1983), Murrindindi (Houghton 1986), Rubicon (Evans 1994), Gembrook (McCarthy, M. J. 1987), west Gippsland (McCarthy, M. J. 1993) and the Wombat Forest (Houghton 1980). These areas lie on and south of the Great Dividing Range, where higher rainfall and better quality forests justified greater investment in tramway operations.

Tramway studies are usually conducted at a district scale, interweaving documentary, photographic and oral sources of evidence. A focus upon industrial aspects is generally evident, such as the routing and construction of tramlines, and locomotive operations. They also provide numerous short histories of sawmilling firms and timber merchants from the late nineteenth and early twentieth century. Tramway studies generally ignore, however, analysis of the social history of mill settlements and site formation processes, and often neglect to place the activities of timber-getters within broader contexts of regional historical development. Nevertheless, such studies provide an invaluable resource in tracing the history and processes of forest exploitation in Victoria.

Forest historians have also traced the origins, development, activities and administration of a number of timber companies which have grown from humble beginnings into large industrial enterprises (e.g. Carter 1986; Graeme-Evans 1996; Mills 1986; Sinclair 1990; Tregenza 1991; Twigg 1994). Major changes to the timber industry from the 1930s onward, in terms of size, location, products, technology and structure, mean that few firms from the early twentieth century have survived to the present day (Dargavel 1988: 109). Several timber companies have commissioned individual histories, celebrating either a centenary of business operations (e.g. Carter 1986; Twigg 1994), or commemorating the firm's demise (Graeme-Evans 1996). Individual companies typically had a close association with particular areas or regions, such as Bunnings in Western Australia (Mills 1986), Bowens in Melbourne (Twigg 1994), Le Messuriers in Port Adelaide (Tregenza 1991) and Risbys in Tasmania (Graeme-Evans 1996). Key elements of these studies include a focus on family members and generational succession, employee loyalty, and the diversification of company interests to compete in a changing commercial environment. Such histories tend, however, to eulogise the companies concerned, and neglect to position them firmly within broader economic contexts. Nevertheless, corporate timber studies complement research into forest industry history and make a useful contribution to Australian commercial history. They trace the development of timber firms through the twentieth century against a background of industry restructuring and rising capital investment, changes in government forestry policy, product marketing, plantation development and pulp processing, and the rise of the modern environmental movement.

The Archaeology of the Twentieth Century

The archaeology of the twentieth century remains poorly documented in Australia. Most research has focused on the colonial period between 1788 and 1900, largely neglecting the last 100 years of social, economic and material development. A similar reluctance to pursue a richer understanding of this period in Britain has been discussed by Matthews (1999; see also Stratton and Trinder 2000) and in the United States by Orser (1999). In Australia, those studies which include consideration of twentieth-century sites or materials typically maintain a primary focus on the later nineteenth century (e.g. Capon 1993; Jack, Holmes & Kerr 1984). Allison's (1998) investigation of the Kinchega Homestead in western New South Wales, and Grimwade's (1998) discussion of the Canning Stock Route in Western Australia are rare examples of archaeological analysis in Australia which concentrate on the twentieth century (see also Marsden 1993; Stuart 1993). More work is contained in consultants' reports but this material is seldom published.

Relatively little is thus known about how most people actually lived in this recent 'familiar' past. Mail order catalogues from the turn of the century reveal the quantity and diversity of goods available, but say nothing of individual responses to such goods, the processes of acquisition or the symbolic content of commodities. The wealth of evidence concerning the recent past, in photos, archives, living memories and archaeological deposits, permits and demands a deeper appreciation of people's daily lives and activities. Much more basic work is still needed to document the manufacture, chronology, distribution and consumption of the many common and utilitarian artefacts typically recovered from later nineteenth and twentieth century sites. It is as important to recover the voices of 'those of little note' from the twentieth century as of any other period. The urban poor, rural communities, itinerant workers and ethnic groups all have stories to tell (Lydon 1999; Marsden 1993: 142-3; Matthews 1999: 158; Orser 1999: 145; Scott 1994).

Archaeologists in Australia have also been reluctant to engage with the explosion of recent scholarship in environmental history. This cross-disciplinary field broadly seeks to document the relationships between people and the environment through time. It encompasses a wide range of themes in Australia, including changes to native vegetation (e.g. Barr & Cary 1992; Benson & Redpath 1997; Bowman 2000; White, M. E. 1994; Williams, M. 1997) perceptions and management of natural resources (e.g. Bonyhady 2000; Hay 2002; Seddon 1994; Wright 1989), climate change (Robin, Sharratt and Griffiths 2005), personal experiences of the bush (e.g. Borschmann 1999; Bunbury 1997; Cubit 1996), forest economics (e.g. Dargavel 1995; Doughty 2000), and the impacts of fire (Pyne 1991) and introduced animals (Coman 1999; Parsonson 1998; see also Dovers 1994, 2000; Flannery 1994; Griffiths 1992, 2001; Griffiths & Robin 1997; Tyrrell 1999). The discipline feeds into contemporary appreciation of the natural world and a concern to repair the damage of the past. This study converges with environmental history by exploring the ways people adapted to living in an isolated forest area. It provides insight into the material lives and cultural adaptations of people living and working on the economic and geographical margins of mainstream Australian life.

Archaeology and History

In their pursuit of the past, historical archaeologists have continually confronted the task of integrating divergent sources of evidence, Schuyler's 'spoken word, written word, observed behaviour and preserved behaviour' (Schuyler 1978b). Some have sought to maintain independent epistemological statuses for archaeological and documentary records, arguing not that documents are subjective and artefacts objective, but that each corresponds to different representations of human behaviour, produced by different people for different purposes (Ferguson 1977: 7; Leone 1988: 33). Others have taken issue with this view, querying whether apparently independent sets of data (documents and artefacts) are genuinely the result of separate historical and social processes (e.g. Johnson 1999: 30; Schuyler 1978a). My view is that artefacts and documents, while qualitatively different, must be critically and imaginatively integrated in the development of plausible, convincing representations of the past. While written records are *intentional* representations of people, places or events, material culture represents the *non-intentional* material by-products of past lives. Archaeological material thus permits evaluation of the unconscious, unconstructed record of the past.

Methodologically, several different orientations are apparent in the treatment by historical archaeologists of written records and material remains. The 'handmaiden to history' approach has used artefacts to reinforce, illustrate and promote conventional historical narratives, often in the support of architectural and historical reconstruction (see Deagan 1982: 158-60; Johnson, M. H. 1999: 29; Little 1994: 7). Artefacts have also served to 'fill in the gaps' in the historical record, providing data on places and technologies poorly represented in documentary sources. This approach has been especially productive in recovering the voices of those who have been ignored, excluded or simply forgotten in historical documents, 'the dispossessed, the downtrodden, and the disenfranchised' (Orser 1999: 145).

Another approach has regarded documentary information and archaeological data as essentially independent and contradictory. The identification of anomalies, ambiguities and disagreements between the contextual sources forces a continual reappraisal of questions and evidence to refine and enliven understanding of the past (Deagan 1988: 8; Leone & Potter 1988: 13-14; Little 1992: 4). Written documents and material remains each provide a particular, largely independent, perspective on historical issues and cultural processes. The tension maintained between them in the process of integration and synthesis provides information and insight of greater depth and detail than if only one set of data was used (Beaudry 1995: 1-4; Deetz 1977: 13). David Lowenthal argues that:

> Memory, history, and relics offer routes to the past best traversed in combination. Each route requires the others for the journey to be significant and credible....Significant apprehension of the past demands engagement with previous experience, one's own and others', along all three routes (Lowenthal 1985: 249).

This study employs elements of all these approaches. Contradictory evidence concerning site formation processes and alcohol consumption, for example, has forced a continual reassessment of questions and sources to achieve a better understanding of cultural life at Henry's Mill. Material evidence relating to living conditions, gardens, foodways, medical responses and notions of respectability has stimulated a range of documentary inquiries to animate and expand these issues. The prevalence of written records concerning teachers, community events and individuals, however, has also added texture to the story of the mill, and demanded a more subtle consideration of physical evidence. Neither documents nor artefacts should be privileged over the other. The unique, simultaneous access enjoyed by historical archaeology to diverse sources of information enables, at least in principle, the pursuit of a richer, more nuanced understanding of past human lives and society.

Background to the Study

Timber-getting was one of the earliest extractive industries conducted in Victoria. Eucalypt forests were widely exploited from the 1830s onward to provide the materials demanded by the growth of European settlement. By the 1850s, steam-powered sawmills began to replace manual cutting, heralding an industry which has flourished in Victoria ever since. Timber products drawn from forests and woodlands included the huts and stockyards of pastoralists, along with shipping timbers, harbour piles and bridgeworks. The rapid expansion of the colonial railway system from the 1860s created an annual demand for hundreds of thousands of hand-hewn timber sleepers. The gold mining industry consumed an immense quantity of shaft timbers and fuel for boilers, while the flood of immigrants brought by the gold discoveries created a boom in housing construction. Timber also provided fuel for brick kilns and bakers' ovens, city paving, vineyard frames, scaffolding, telegraph poles, carts and carriages, barrels and boxes, and immense quantities of firewood.

Sawmills were often located in the midst of forest areas, as it was generally easier to move a mill to the trees than trees to the mill. Their nature, scale and longevity of operation varied considerably in later nineteenth and early twentieth-century Victoria. Ephemeral 'spot' mills, for example, typically operated for only a season or two, using a single saw bench and a small crew to produce timber for local needs. They were often employed to clear a back paddock on privately held land. Larger mills were characterised by higher motive power, greater timber output and bigger crews, generally operating on larger forest areas. Most forest-based mills were nevertheless erected as temporary operations, only designed to last until the available timber was cut out. In the Otways, for example, the average length of operation between 1852 and 1939 was around five and a half years (Brinkman & Farrell 1987: 175-9). Mills relying on horses or bullocks to haul logs and timber tended to operate seasonally, as tracks became boggy quagmires in winter months. The use of timber tramlines, however, connecting sawmills to railways, permitted the exploitation of previously inaccessible forest areas and the year-round operation of mills (Plate 2). The dominance of railways was nevertheless gradually eroded during the 1930s, as motor trucks began transporting greater quantities of logs and timber. This eventually removed the imperative for sawmills to be located in the midst of forests at all. The wildfires of 1939, the impacts of the Second World War and the shifting of mills into towns rendered the old ways of life in the bush obsolete.

Residential patterns at sawmills between the 1850s and 1930s varied according to the nature of forest resources and the geographical proximity of mills to nearby towns. In some cases men travelled out to bush mills from town on a daily basis, returning home in the evening. Elsewhere they worked on site during the week, camping in huts or tents, and returned to their families on weekends. In other cases workers lived permanently at the mill in small wooden huts, with a boarding house to provide meals. At larger, more isolated mills, substantial settlements were often created when married men brought their families to live on site. Such camps were widely distributed throughout the timber-producing areas of Victoria during this period. The establishment of houses and gardens promoted a sense of community life, while a school and store helped to meet the basic social and material needs of resident women and children. Such mill settlements often took on the appearance of small townships, but with an over-riding focus on sawmilling. Henry's No.1 Mill was typical of such settlements.

Regional biophysiography

The site of Henry's No.1 Mill today is located in the Great Otway National Park under the management of the Department of Sustainability and Environment. The site extends about 400 metres along the south bank of the West Barwon River, 230 metres above sea level. Rainfall in the vicinity of the No.1 Mill is in the order of 1400-1600 mm per annum. The mill lies in a zone of riparian wet sclerophyll forest, dominated by an overstorey of manna gum (*Eucalyptus viminalis*) and an understorey of smaller trees, ferns, herbs and climbers. The site itself represents a mosaic of forest regrowth, flood zone scrub and introduced grasses and shrubs, retarding easy movement and severely restricting surface visibility. Vegetation along the river zone consists of a mixture of indigenous and exotic, opportunistic species, probably due

Plate 2. Henry's Barwon tramline

to the high frequency of disturbance by flooding in this area. Manna gum and blackwood (*Acacia melanoxylon*) regrowth, diverse shrubs and tree ferns, and herb-layer brackens and nettles dominate the residential part of the site and gully areas.

The geology of the Otway region is represented mainly by a parent material of non-marine sediments, laid down in the Lower Cretaceous period over Paleozoic basement rocks. The Otway Ranges consist of these Mesozoic sediments uplifted into a complex domal structure. The range extends for almost 80 kilometres in a generally south-westward direction from Peters Hill in the east to Olangolah, where it then forks. One branch of the ridge turns in a southerly direction to Cape Otway, while the main ridge continues south-west through Beech Forest to Lavers Hill. The highest point is Mount Cowley (670 metres), and the elevation of the main ridge is about 500 metres for much of its length. Dissection of the range is deep, with a radial stream pattern reflecting the dome-like structural control. The headwaters of the Barwon River in the east and the Gellibrand River in the west flow from the northern slopes of the range. Many short, swift-flowing rivers drain the steep southern slopes, including the Aire, Parker, Barham, Kennett, Wye, Cumberland and Erskine Rivers. Waterfalls (e.g. Erskine, Cumberland, Triplet Falls) are prominent where these streams cross hard sandstone bands, or where faulting has resulted in prominent scarps. Landslips are also a feature of the region (Douglas 1977).

The Otway Ranges exert a significant influence on local rainfall and temperature, in comparison with the coastal area to the south and the plains to the north. Although the bulk of rain falls in April-November, effective rainfall in the ranges does not normally limit plant growth at any time during the year. Average annual rainfall measured at Weeaproinah in the western Otways exceeds 1900 mm, while in wet years annual totals have exceeded 2700 mm, ranking among the wettest areas in Victoria. Mean monthly temperatures in the Otways range from 11-21° C in February, down to 4-8° C in July. Fog and frosts occur most commonly in winter, generally dissipating a few hours after sunrise. Snow occurs on the heights of the main Otways ridge two or three days per year, and has fallen to depths of five to ten centimetres (Brinkman & Farrell 1990: 3-6; Linforth 1977).

The vegetation structure of the Otways varies considerably, reflecting the diversity of such factors as geology, soils and climate. Variations in vegetation communities of the area are characterised by the shifts and terminations associated with the Tertiary-Lower Cretaceous boundary, and the gradual change in species structure corresponding to the availability of soil moisture. The Cretaceous rocks which comprise the Otway Range have weathered to form moderately fertile soils with loamy topsoils. Tertiary sediments to the north generally produce highly infertile soils with sandy topsoils (Parsons et al. 1977: 77). Tall wet forests, with stands up to and exceeding 40 metres in height, thus dominate the Otway Ranges. The stands flanking these ranges, where rainfall and soil fertility is lower, carry shorter, drier forests, heaths and woodlands. These vegetation profiles contrast with the remnants of natural grassland on the basaltic plains to the north, and the low coastal scrub and grasslands on coastal dunes to the south. The Cretaceous core of the Otways thus exhibits a moisture-related continuum of eucalypt communities. Red ironbark (*E. sideroxylon*), narrow-leaf peppermint (*E. radiata*) and scent-bark (*E. aromophloia*) are confined to the drier areas, while manna gum (*E. viminalis*), blue gum (*E. globulus*), mountain grey gum (*E. cypellocarpa*) and messmate (*E. obliqua*) form a central set in the wetter areas, along with blackwood (*Acacia melanoxylon*), myrtle beech (*Nothofagus cunninghamii*) and satinwood (*Phebalium squameum*). Mountain ash (*E. regnans*) forms pure stands above 490 metres and mixed stands below this elevation in the southern Otways (Parsons et al. 1977; Williams, L. 1977).

The vegetation profile, topography and climate of the Otways all played crucial roles in the development of the region's timber industry. The high density of large trees in the upper ranges meant that numerous mills operated in the area between 1900 and 1930, contributing a significant proportion of the State's timber output. Steep terrain and the quality of forest resources also justified and frequently necessitated the construction of steam-powered winches to haul logs, as rivers and gorges were too steep and rocky to be used for timber transport. High rainfall meant that roads and tracks were impassable for much of the year. Tramlines were thus generally needed to transport people, equipment, food and timber. In spite of the removal of forest mills to towns after World War II, logging has continued in the Otways to the present day, generating controversy over harvesting quotas and export woodchips.

Sources used in the Research

The following section outlines the main sources and forms of evidence utilised in the present study, including oral history, photographs and historical documents. Material culture, in the form of excavated assemblages and artefact analyses, is presented at length in Chapters 3 and 4. The aim here is to provide an indication of the scale and quality of historical source material available for a detailed study of community at Henry's No.1 Mill. Due to its size and length of operation, there was greater opportunity for records concerning the site to accumulate than for other mills operating during the same period. Government authorities involved in forest management, education and postal services had formal interests in the mill which created official records. As one of the largest timber mills in the region, Henry's also attracted newspaper attention on a scale commensurate with a small township, while occasionally the misdeeds of its residents demanded the attention of law courts. Residents also commonly cited 'Henry & Son' as their address for electoral enrolment.

The approach I have taken to ordering this information derives in part from Darrett and Anita Rutman's historical ethnography of Middlesex County, Virginia (1984). Their method of 'record stripping' disassembled all of the records pertaining to the place and reordered them on a name basis to create their own census of the county. In the present study, birth records, school correspondence, newspaper accounts, electoral rolls and other materials have all contributed to constructing a biography of the mill through its people. The resulting database includes a field for each mill resident, so that people could be tracked, either as individuals or groups, according to sex, occupation, marital status, kin relations, and period and length of residence (Appendix B).

The school at Henry's Mill operated from 1909 till 1927. The Victorian government introduced free, secular and compulsory education in 1872, establishing an Education Department to employ teachers and regulate schools. Department officials maintained a voluminous correspondence with teachers in each of the several thousand primary [elementary] schools scattered across Victoria. The resulting Education Department files provide an especially rich source of information on the school at Henry's Mill, and its teachers and pupils. The two most useful series of records identified were the Primary Schools Correspondence Files (1878-1962) and the School Building Files (1868-1975). The former consists of letters written by teachers to the department concerning transfers, problem students, teacher examinations and general school conduct. The latter series comprises letters regarding the establishment and maintenance of the school building and its fittings. The corresponding departmental officer annotated most documents as a prelude to formal typed replies. The processes of comment, complaint and response to school problems can thus be identified by close analysis of these documents. Although state schools were notorious for not retaining class enrolment lists, attendance levels were occasionally referred to by various teachers at Henry's, revealing a gradual increase in student numbers. Useful comparative material is also available from other forest mill schools, including Mississippi Saw Mills, Upper Britannia Creek and Rubicon.

It is ironic that in an industry typically characterised as male-dominated, one of the best sources of evidence on mill life derives from the presence of children and the need to provide for their schooling. Other groups, especially single men, are poorly represented in the historical record. The documentary bias against unmarried workers is reinforced by the paucity of archaeological remains representing their lifeways. This presents a challenge to understanding this segment of the mill population, and to the stereotype of their domination of the industry. Analysis of community is thus skewed towards the material and social life of married men and their families.

From the nineteenth century up to the establishment of the Forests Commission (FCV) in 1919, Victoria's forests were managed by a succession of government departments, including Crown Lands and Survey, Agriculture, Mines and, from 1908, the State Forests Department (FCV 1928: 41). Constituted under the *Forests Act* 1918, the FCV was responsible for managing indigenous forests, establishing softwood plantations, silvicultural research and fire prevention. The Commission's work rapidly created large volumes of correspondence, organised via an intricate registration system (McRae 1994). However, 'Prior to 1920 files' inherited by the Commission upon its formation which dealt with finalised matters were eventually destroyed (McRae 1994: 12, 27-8). Early correspondence from Henry appears to have met this fate. This means that only one file of correspondence between W. R. Henry & Son and the FCV is extant (FCV 29/0474). Preserved files from Clark & Pearce (Rubicon), Sanderson (Barramunga) and Hayden Brothers (Barwon Downs) provide important comparative material broadly contemporary with Henry's operations. Early business records from the Henry family, however, have not been preserved (R. Henry 12 June 1999).

Registers of district Courts of Petty Sessions were also consulted to gain insight into crimes and misconduct committed in the area. Such courts were established and convened regularly at Beech Forest and Birregurra, the former dealing with court matters in the western Otways, the latter covering the eastern Otways. The great majority of cases to come before a magistrate involved such issues as debt recovery, petty theft, public drunkenness and obscene language, and the failure of parents to ensure the school attendance and medical vaccination of their children. Proceedings of greater seriousness were generally referred to higher courts in Geelong. In some cases, complainants and defendants from Henry's Mill were identifiable from electoral registration and other sources.

Electoral rolls were prepared annually for the Commonwealth Government from 1901. Women gained the right to vote in federal elections in 1903 and in Victorian state elections in 1908 (Summers 1994: 393). Voting was not compulsory, however, until 1924, and only those 21 years of age or more were eligible. The rolls consist of an alphabetical register of names, along with the address, occupation and sex of each voter. They are structured by divisions and subdivisions, with Henry's Mill located in the Division of Corangamite, in the Subdivision of Forrest. The mill was one of seven polling stations in the area, due to its size and geographical position. Electoral rolls were combed for every available year of the mill's operation, to measure population change and the length of time people resided at the mill. The aim was to develop a community profile in terms of occupation, sex and marital status.

District newspapers shed valuable light not only on people and events at Henry's Mill, but also on the broader economic and cultural context of the region. Newspapers available for examination include *The Birregurra Times* (1918-22), *The Winchelsea Shire Advertiser* (1910-11), *The Colac Herald* (1902-28) and *The Colac Reformer* (1902-20). A monthly newspaper, *The Timber Worker*, was produced by the Victorian branch of the Amalgamated Timber Workers' Union of Australia to maintain contact with members. It regularly included information on Arbitration Court proceedings, wage determinations and industry conditions.

Although a post office operated at the mill, few records survive from the Postmaster-General's Department relating to the early twentieth century due to First World War paper shortages and requisitions (Davies 2005b; CH 23 July 1918). Australian Archives preserve only skeletal indexes of post office correspondence. Similarly, census records are sparse, as it has long been government policy to destroy manuscript census returns 'to protect privacy' once vital statistics have been extracted (Vine Hall 1994: 73). Local shire rate books also have limited potential in this study, because workers at Henry's did not own residential property at the mill. The settlement was erected on Crown land under the terms of a three acre sawmill license issued to W. R. Henry & Son. Mill sites were not surveyed and subdivided, so that no land title or other spatial information is available. As families rented homes from the proprietors, there is no official record of the succession of residence at each mill house. Rate books from the Shires of Colac, Winchelsea and South Barwon were still examined, however, to determine the extent to which residents at the No.1 Mill acquired rateable property elsewhere in the district. The results were meagre, suggesting that few people from the mill settled in the area for substantial lengths of time. This is reinforced by cemetery records, which indicate that only six people associated with the No.1 Mill were ever buried at the nearby Yaugher (Forrest) cemetery.

Memoirs are another source of information on the social history of the timber industry. Usually written by former mill workers, they provide an eclectic array of anecdotes, memories and yarns, frequently leavened with droll stories and poems. Although largely unverifiable and typically devoid of references, the freshness and immediacy of their accounts and the direct contact enjoyed by the authors with former mills and workers adds value to what would otherwise be an unreliable source of evidence. Jack Loney, for example, has prepared numerous publications on the Otway region and the shipwrecks along its treacherous coastline (e.g. Loney 1979, 1989). George Facey's accounts concentrate more on the timber mills and pioneer settlers of the Otways (e.g. Facey, G. 1977, 1981; see also Mackie 1980), while Corker Brown's 'The Otways that I knew' offers a fairly substantial treatment of the region and its historical development (Brown, C. 1993).

Memories of informants were also sought to provide depth and detail in understanding patterns of work and everyday living, games and pastimes, personalities and social events. Their information supplemented, challenged and widened evidence from other sources on housing and health, diet and drink, and responses to the environment, as well as the roles and activities of children. Interviews were conducted with seven people, either in person or by telephone. These included five men (Jim Biddle, George Chamberlain, Norm Gardner, Ric Henry and Jim Washington) and two women (Elsie Chamberlain and Joyce Harding). The only informant to have first-hand knowledge of the No.1 Mill was Ric Henry, grandson of the proprietor, W. R. Henry. Born in 1917, Ric spent his school holidays at the mill during the 1920s with his parents and two older brothers, Fred and Ivan. Besides Ric Henry, however, all the informants were associated with Henry's later mills, especially the Nettle (1937-43) and Carisbrook Creek sites (1938-48). Jim Washington, for example, born in 1921, worked as a faller and tallyman at the Carisbrook Mill in 1941. Jim Biddle was 18 years old when he started work at the same site around 1945, and remained at the mill for the following four years. Joyce Harding was born in Melbourne in 1920. Arriving as a newly-wed after the 1939 fires, she and her husband Leo spent two years at the Nettle Mill. Norm Gardner worked at the Roadknight and Fun Factory Mills as a young man in the 1930s. Elsie and George Chamberlain arrived at the Carisbrook straight after their wedding in 1938, and spent the next six years at the mill. The young age of these people during their time at the mills reinforces the youthfulness of many workers in the timber industry during this era, but also reflects the fact that older workers passed away before this study began. Although employed one or two generations later than the period of the No.1 Mill's operation, their memories provide vivid insight into the material culture and living conditions at Henry's mills beyond the Otway divide.

Significant photographic records of forest-based industries in late nineteenth and early twentieth century Victoria include those of the State Library of Victoria and the National Library in Canberra, along with light railway and timber tramway histories. Photos have been regarded here less as illustrative supplements than as a means to assess general aspects of mill life, including timber-getting, transformations of forest landscapes and the composition and size of mill crews. The images typically reproduced in light rail histories depict the transport technology of tramways and locomotives, groups of timber workers, and the logs and sawn timber they produced. There are generally few examples, however, of mill settlements, housing and amenities, or of the women and children who formed such important components of forest mill communities. Such exclusion betrays the bias of both photographers and tramway historians, in their concentration on the technology of the timber industry, at the expense of its social and cultural elements. The focus on mill crews, however, helps to balance the absence of single men in documentary and archival sources.

Outline of the rest of the Study

The following section provides an outline of each chapter in the present study, summarising the central concerns and arguments presented. The historical, industrial and economic context of forest-based sawmilling in Victoria is reviewed to begin with, providing the context for the detailed investigation of Henry's Mill. Chapter 2 thus begins with a brief overview of the Victorian timber industry from the late nineteenth century until the Second World War, describing the economic fluctuations of the industry, logging districts, forest products and changing management policies. A detailed summary of sawmilling technology and operations is then provided, demonstrating the labour-intensive nature of the industry, its reliance on steam and animal motive power, and the dispersed, isolated nature of forest-based sawmills and settlements. An outline history of timber-getting in the Otways is then provided. Exploitation of the magnificent forests of the region had commenced as early as 1849, but expansion of the industry was hampered by the lack of suitable transport infrastructure. Completion of the Birregurra-Forrest railway in 1891, and the narrow-gauge line from Colac to Beech Forest in 1902, however, stimulated a boom in Otways timber production in the following years. Sawmills were set up throughout the region, linked by timber tramways to the new rail lines. By the 1920s and 1930s, roads and motor trucks were gradually replacing horses, bullocks and tramways, with the consolidation of many smaller forest mills into fewer, larger town mills from the 1940s onward. Henry's No.1 Mill thus operated during the heyday of the Otways timber industry. Although larger and longer established than most, it typified the nature of forest mills and settlements of the era. The mill's proprietors were nevertheless prepared to risk capital and apply innovation in the pursuit of timber resources and higher profits. A tramline tunnel, for example, 290 metres in length, was the first of its kind in Victoria, while the mill was one of the earliest in the region to adopt steam locomotives and winches. Additional, smaller mills were established by W. R. Henry & Son in 1911, 1920, 1921 and 1925. The destruction of the No.1 Mill by fire in 1927, however, resulted in the gradual withdrawal of timber-cutting from the West Barwon valley. Valuable plant and equipment was removed, workers and residents salvaged everything possible from their dwellings, before final abandonment of the valley early in 1929.

Following this background history of forest operations, Chapter 3 presents the field methodology and archaeological results of site survey and excavation at Henry's Mill. Survey aimed to create a base plan of the main physical and cultural features of the site. Subsequent excavation concentrated on three house sites, Trenches A, B and E, along with a sounding in the area of the single men's huts (Trench D) and an intensive surface collection in the discard zone along the river flat below the tramline (Trench C). Detailed descriptions of excavation results for each trench are followed by assessment of site abandonment processes and their impact on the survival of cultural material.

Chapter 4 presents analysis of artefacts recovered during excavation, characterising the assemblage in terms of its scale, range, and diversity of forms and fabrics. A functional classification system has been employed, consisting of five broad activity sets (architectural, industrial, domestic, personal and social), each subdivided into several functional classes. The aim has been to isolate specific data sets and facilitate their analysis in relation to such issues as domestic consumption patterns, health, hygiene and environmental responses. Among the 100+ artefact forms identified, represented by more than twelve thousand artefact fragments, the most numerically significant items were nails, glass bottles and ceramic tablewares.

Specific dimensions of community life are examined within the following three chapters, which draw on and integrate the information already presented. Chapters 5-7 concern the spaces and structures of the site, the people who lived and worked within them, and the sense of belonging generated among residents. Chapter 5 examines the architecture and layout of the mill settlement, which provided the physical context for expressions of community. The mill shed and its associated yards and tramlines formed the industrial core of the site, occupying a large proportion of the available level terrain. Amenities such as the boarding house, post office, store and billiard room were located close by, while the small huts of single men clustered between the tramlines and the base of a steep spur to the west (Figure 4). Houses for married workers with resident families, however, were located 150-200 metres to the north-west in a discrete 'suburb' known as 'Port Arthur'. This separation was typical of sawmill settlements of the period, reinforcing in concrete terms the social distinction based on marital status. Construction of the school building within the midst of these household dwellings emphasized their orientation towards family life. Excavation of three family house sites revealed that each house had two separate hearth features: an open fireplace in the main room and a bricked-in iron oven at the rear. These remains hint at the isolation of women in rear kitchens while their families gathered around the main open hearth in larger front rooms. The spaces within and between the architecture of the site both expressed and magnified the industrial, social and gendered divisions within the mill community.

The following chapter seeks to characterise the people of Henry's Mill, to examine who they were and from where they came. This involves assessment of the size and structure of families at the mill, the position of unmarried male workers, and migration patterns to and from the site. Mill proprietors generally preferred to employ married workers, regarding them as more stable and committed to their employment. Children thus formed a prominent part of the mill population, with a mode of three per family. The number of families at the site nevertheless varied over time between about 10 and 20, generally with 30 or 40 single men on the payroll as well. Attention is also paid to the schoolteachers at Henry's. Occupying a distinctive role within the community, and separated by their training and professional authority from other mill residents, their concerns and complaints provide vivid insight into life at the mill. The lives of certain individuals are then traced, to humanise the abstraction of numbers. Attention focuses on those whose individual qualities, general circumstances or personal adversity permit their 'excavation', or retrieval, from living memory or documentary sources. These include Gus and Bernie Seebeck, Gladys Butcher and her family, Cecil Turner, Frank McGlone and his brothers, and W. R. Henry and his son and partner, Ern. Together, these individual case histories provide a coarse cross-section of the mill community, and offer more general understanding of the hard work, personal expectations and community conflicts experienced by rural workers in Australia during the early twentieth century.

Chapter 7 examines the notion of community at Henry's Mill. The psychological element of community involved a sense of belonging, identification with a place and its people, and their aspirations, fears and concerns. One of the most important expressions of community was thus through stability of residence and the continual reinforcement of family, neighbourhood and friendship ties. The high turnover of workers at Henry's, however, meant that a sense of belonging was limited. It was rare for an employee to remain at the site for more than three or four years, which resulted in the constant arrival and departure of individuals and households. Sectarian and political divisions also conditioned the sense of place, as did the distinctions recognised in employment status. Although occasions such as company picnics, sports days and school fetes were unifying occasions for mill residents, their sense of attachment and identification with the community was always qualified by the mobility of the workforce and by networks of exclusion and collaboration.

By the early twentieth century, even very isolated rural settlements routinely acquired industrial goods, typically incorporating them within local frameworks of meaning. A wide array of artefacts from Henry's Mill is used in Chapter 8 to examine the nature of subsistence and consumption practices at the site. The manufacturing origins of 71 separate items provide evidence of the interaction spheres in which residents at the mill participated. It is evident that goods from local, regional, national and international markets were available to consumers at the site. Although the United States was represented by only three patent medicine bottles, Great Britain was an important source of alcohol, condiments, ammunition, tablewares and other commodities. The bulk of identified manufactured items, however, derived from Melbourne, and included bricks, medicines, ironmongery, food and beverages. Further supplies originated in Geelong, Ballarat, Adelaide and Sydney.

Workers and residents at the No.1 Mill acquired these goods through various mechanisms, which are also examined in this chapter. Most importantly, the timber tramline linked the mill with the railhead at Forrest and the world beyond. Two general stores in the township served the immediate needs of the local district, while more extensive facilities were available in the larger towns of Birregurra and Colac. The Henrys also operated a small store on site, providing mostly small items for the unmarried men, while hawkers may have made occasional visits to the mill as well. In an era of mass production and consumption, however, the people at Henry's stretched their salaries and avoided complete reliance on the cash purchase of goods by growing vegetables and occasionally catching rabbits and fish. Materials such as bricks, kerosene tins, glass bottles and machinery components were also scavenged, reused and recycled. Residents were thus integrated with mass consumer culture, while their geographical remoteness and limited incomes intensified bush traditions of resourcefulness and improvisation.

Attention also focuses on specific elements of consumption and their meanings within the mill community, integrating documentary information and various archaeological data sets. Alcohol, for example, is conventionally regarded as playing a central role in the lives of mill workers, especially in the context of 'spree' drinking, but material evidence from Henry's suggests a more moderate intake of alcohol and a more responsible attitude to drinking and worker safety. The clothing worn by men, women and children was typically durable and inexpensive, enlivened occasionally by a brooch, button or accessory which discretely reworked everyday attire into something more respectable (Lawrence 2000: 129). Domestic interiors of the rough bush dwellings were enhanced by select items such as pressed glasswares and pictures, promoting a modicum of comfort and reinforcing notions of decency.

The theme of isolation and integration is developed further in the next chapter, which explores responses to injury and illness within the mill community. In an era when professional medicine played an increasingly important role in public health, the remoteness of the No.1 Mill from hospitals and doctors forced residents to maintain alternative medical treatments. The range of injuries to which workers were subject is thus reviewed, along with the general living conditions, climate and quality of water supply as factors affecting the health of women, children and men at the site. The development of professional medical practice by 1900 from earlier, rudimentary regimes provides the background to a more detailed assessment of the responses adopted by mill residents to their health and medical needs.

Doctors were distant, expensive and only consulted in emergencies. Bush Nurses provided a more local alternative, offering advice to nursing mothers and making regular visits to district schools. Death and disability insurance was available to members of the Timber Workers' Union, while informal 'pass the hat' strategies offered additional, informal support mechanisms to injured workers. Material evidence from the No.1 Mill also indicates that various non-prescription remedies were employed to combat illness. Homeopathic preparations were identified from the remains of small glass phials and bottles, while at least 35 patent medicine bottles were also recovered, many of which were marketed as cures for bronchial infections and children's ailments. Castor oil and kerosene were used to treat colds and sore throats, while a small plastic ring may be interpreted as a contraceptive pessary or a teething ring. Self-diagnosis and treatment thus appears to have been widely practised at the site. Residents used commercial preparations to overcome the health problems posed by living and working in a remote mountain forest environment.

The final chapter reviews the aims and conclusions drawn from the present study. The No.1 Mill was one of thousands of ephemeral settlements which made up the rural landscape of this period. Its detailed investigation adds significantly to understanding of twentieth-century archaeology and rural community life in Australia. The integration here of archaeological material with historical and oral sources of evidence aims to deepen understanding of daily life at forest mill camps. The work was dirty, difficult and dangerous, and living conditions were basic. Residents constantly struggled to overcome their physical isolation by engagement with the world beyond the forest. A wide array of consumer goods were acquired, used and discarded, although bush traditions of improvisation and thrift persisted.

Once common throughout the forest areas of Victoria, bush sawmills and the way of life and work they represented have passed into history. The timber industry nevertheless played a vital role in local and regional economies, providing forest products for use in a vast array of domestic, commercial and industrial contexts. Timber tramlines opened up previously inaccessible areas, providing vital year-round transport and communication infrastructure into remote mountain forest environments. Sawmills employed thousands of men, providing jobs for struggling farmers and their sons, rural labourers and skilled tradesmen. The frequent movement of men and their families in and out of the industry was part of the mobile, migratory nature of much rural work of this time. Timber workers created settlements, generally small and short-lived, which nevertheless were home to men, women and children for months or years on end.

Chapter 2

Historical Context

Although W. R. Henry and Son operated in a pysically isolated location, their activities were part of a broader economy of forest exploitation. The work practices and technology they utilised, and markets for which they competed, were similar throughout the timber industry. The following discussion thus begins by outlining the historical context of timber-getting in nineteenth and early twentieth-century Victoria. It then reviews historical changes in forest policy and management, and the processes involved in transforming forest trees into timber. The fluctuations, products and technologies which characterised the industry during this period provided opportunities and challenges for sawmillers in the Otway Ranges. A review of their activities establishes the regional framework within which W. R. Henry and Son came to be prominent figures, and contextualises the experiences of those who lived and worked at their mills. The most significant elements of the Henrys' success involved access to the substantial forest resources of the West Barwon valley, the innovative use of available technology, maintenance of a large workforce, extensive use of timber tramlines, and successful negotiation with government authorities over cutting areas and royalty rates.

Timber industry in Victoria

Timber-getting was a vital activity from the beginning of permanent European settlement in Victoria in 1835. Forests and woodlands provided material for huts and houses, fuel for domestic fires, as well as barrels, boxes, carts and wheels. The region was blessed with an abundance of forests and woodlands which sawmillers gradually learned to exploit. Large supplies of sawn timber also arrived from Tasmania and New Zealand. While the earliest local timber was cut by hand over a sawpit, the first steam sawmill was erected in Melbourne early in 1841 by Alison and Knight, near the corner of Collins and King Streets. John Manton followed a month or two later, with a mill near Queens Wharf on Flinders Street (Houghton 1982). Fine stands of timber at Mount Macedon prompted William Robertson to construct a water-driven sawmill in the area around 1844 (Hutton 1988).

The discovery of gold in Victoria in 1851 changed everything. It had a sudden and massive impact on the local economy, society and environment. Goldseekers arrived by the tens of thousands, creating a demand for all kinds of timber products, including housing, fencing, fuel and furniture. The mining industry itself was also a voracious consumer of timber for pit props, boiler fuel, poppet legs and building construction. By the 1860s, the timber needs of the 85 000 miners working on the goldfields were rapidly destroying surrounding forests. Sawmillers hurried to supply the demand, creating an industrial relationship of mutual dependency. By the end of the 1860s, Ballarat consumed nearly 180 000 tons of firewood, 850 000 props, three million laths and 7.5 million super feet of sawn timber each year. In 1870, these figures had almost doubled (Dingle 1984: 101; Mueller 1871: 10-11).

Gold mining itself also had a dramatic and lasting impact on forest areas. Many of the goldfields were located in wooded hills, ranges and valleys which had previously been little used or ignored. Miners quickly felled huge numbers of trees on and around the goldfields. William Howitt expressed amazement at how the diggers cut trees to provide bark for the roofs and walls of their huts, and timber for posts, rails and slabs: 'No sooner have they done their day's work than they commence felling trees'. Trees were wasted in their thousands, so that 'a man chops down a tree to boil a can of tea' (Howitt 1855: 98). Forests were savaged 'as if the loftiest tree were the growth of a single night' (Brough Smyth 1869: 28). Historian Tony Dingle notes that in England, only the wealthy owned trees and could order them felled, but no effective restriction existed on the goldfields, and 'many diggers revelled in this new-found pleasure' (Dingle 1984: 50). Leaving the ground covered with bark, debris and dead wood increased the risk and destructiveness of fires.

The number of sawmills in Victoria increased steadily following the discovery of gold. There were 30 in operation by 1855, 64 by 1860, and 127 by 1870. The 1880s building boom resulted in 323 sawmills by 1889. The 'bust' which followed, however, saw the number of sawmills reduced to 166 by the end of the century (Dargavel and Boutland 1988). By the beginning of the First World War, there were 378 forest and metropolitan sawmwills operating in Victoria. These directly employed almost 6000 men, accounting for 1.4% of the adult male workforce. There were also thousands more who worked cutting firewood, railway sleepers, fence posts, shingles and palings (Laughton 1915-16: 786-95).

Sawmills in Victoria were generally quite small, compared to the much larger, more mechanised operations in North America. Domestic markets for local hardwoods were limited, and competition from imported softwoods was intense. Patchy forest areas meant it was often better to move a small mill regularly, rather than supply large logs over long distances through rugged terrain. Forest sawmills were thus small, often individual or family-run operations, with a limited amount of capital and machinery. Many of these mills were worked only seasonally, in between other demands of the rural year, such as shearing, harvesting and fencing. Larger sawmills in towns, on the other hand, were built mostly by merchants or builders to supply the metropolitan market. They generally processed both Australian and imported timbers, and often had dressing, moulding or joinery facilities as well (Dargavel and Boutland 1988: 40).

Melbourne and Bendigo were linked by rail in 1862. The line was extended to Echuca by 1864, thus opening up the vast pastoral hinterland of the Murray, Murrumbidgee and Darling rivers to trade with Melbourne and beyond. By mid-1865, 26 steamers were at work on the river system, carrying enormous loads of wool and wood. River red gum (*E. camaldulensis*) was in great demand for railway sleepers, piers and wharves, and mining timbers. One large tree could produce up to 300 sleepers (Bolton 1992: 45). In addition, much of the 'waste' from milling red gum was dispatched to settlers on the treeless plains of New South Wales for building slabs.

The red gum industry was highly seasonal, with logs cut in the forest in summer and autumn. These were then either sawn on the spot by portable steam-powered mills, or hauled to the banks of the Murray by horse and bullock teams. When the river flooded in winter and spring, barges were used to transport logs to large mills in Echuca. By the late 1870s, the red gum trade was reputed to employ up to 1500 men. In the following years, however, over-cutting reduced the available timber and the industry went into decline (Fahey 1987).

By the turn of the century, the major products from Victorian forests were construction and fencing timbers, beams and poles, firewood, railway sleepers, charcoal and eucalyptus oil. Demand for mining timbers and fuel eased in the years prior to 1914, with a lull in the development of the main quartz and alluvial goldfields (Mackay 1908, 1914). The continued expansion of state railways, however, demanded around half a million sleepers annually for construction and maintenance. Contracts for sleepers were restricted to the slow-growing red ironbark, grey box and red gum forests and woodlands. Limited supply of these trees, however, and excessive waste from each log by sleeper hewers, had resulted in a reduction in supplies, and calls to utilise the more common and widespread, but less durable, messmate, stringybark and blue gum (Mackay 1909).

By the 1920s the forests of the Upper Yarra began to emerge as Victoria's premier logging district, accounting for more than half of the State's sawn timber output. The first sawmills in the area were built in the 1890s, with interest in the magnificent mountain ash and messmate forests of the area stimulated by completion of the railway from Lilydale to Warburton in 1901. By 1926 there were 42 mills operating in the district (Carroll 1988: 61-7). Further south, construction of a narrow gauge railway to Gembrook in 1900 brought the area's high quality forest into production for the nearby Melbourne market (McCarthy 1987). West Gippsland continued as a major sawmilling region into the twentieth century, while yellow stringybark from South Gippsland provided large timbers for bridges, wharves and other engineering works.

Sawmillers in the Otway Ranges also contributed a substantial portion of Victoria's timber supply, following construction of several branch lines into the area in the 1890s and early 1900s (Houghton 1975, 1992). The Wombat Forest, straddling the Dividing Range through Central Victoria, had been excessively cut in previous decades to meet the demands of the mining industry, but it continued to be an important source of messmate and other mixed eucalypt timber in this period (Houghton 1980: 12-13). Sawmilling in the Rubicon forest, in the Central Highlands north-east of Melbourne, expanded rapidly following the arrival of the railway at Alexandra in 1909 (Evans 1994: 1-4). Red gum sawmilling also continued in a smaller way in the Barmah Forest, producing mill logs, wood fuel and railway sleepers.

Until the 1890s, a system of 'area' licences was applied to timber-getting in Victoria. The area system required the sawmiller to pay a fixed sum for sole rights to an area from which he could take all the timber he wanted. Beginning in 1893, however, a system of royalty payments was gradually introduced. This reduced the waste of timber that characterised the industry in the nineteenth century, and ensured a much higher revenue for government. By 1902, royalties applied to all Victorian State Forests, with sawmillers paying a set amount 'over the saw' for each piece of timber cut. By 1920, royalties were payable on sawn timber, piles and beams, cooperage stock, sleepers and mining timbers, along with charcoal and eucalyptus oil. Rates applied to three classes of forest produce, based on the quality of the available resource and the distance between forest and market (*Forests Act* 1918: sections 16 and 18). It was not until the 1930s, however, that royalties were paid 'in the round', on logs removed from the forest, to force greater efficiency in log conversion and timber utilisation (Evans 1994: 117).

Fluctuations in demand for timber in Victoria continued during the twentieth century. Recovery following the 1890s depression was halted by the onset of the First World War and a reduction in building activity. The industry briefly returned to full production to meet the demand for housing from returning soldiers, but an economic downturn in 1922 and a flood of imports depressed production again (Dargavel 1995: 50). The combined effects of the 1929 Timber Strike, where city mill workers tried unsucessfully to defend the 44-hour working week, granted under a federal award in 1920, and the onset of the Great Depression, sent demand for timber plummeting again, and forced many mills to close (FCV 1931). Partial recovery of the industry by 1939 was massively disrupted by the Black Friday bushfires of 1939, which saw 69 mills burned down, 71 lives lost, and millions of acres of forest destroyed or badly damaged (Pyne 1991: 309-14).

A Royal Commission on the fires recommended that for safety reasons mills in many areas no longer be located in the forest (Stretton 1939: 34). In the immediate post-war years, the Forests Commission argued that timber workers and their families suffered economic, physical and social disadvantages from living in mill settlements, and that it was difficult to maintain labour, log supplies and transport facilities for a large number of isolated and independent operations (Carron 1985: 190). Fewer workers were prepared to tolerate the rough conditions typical of isolated sawmilling settlements, and most wanted better access to community facilities and education for their children. Electricity provided greater convenience and control for sawmillers than steam engines, so that the extension of the electricity grid throughout Victoria after the Second World War enabled the relocation of mills to townships at the forest edge. This process was also facilitated by the construction of road networks and diesel engine technology. The age of tramways and steam engines, of small bush mills and settlements scattered throughout the forest was at an end, replaced by roads and trucks, electricity and pulp mills, centralised populations and an entirely new industrial economy of forest exploitation.

Managing the forests

Government controls on early timber-cutting in Victoria were minimal and largely ineffective. There was a widespread view that timber should be available to everyone at little or no cost. Demand for grazing land by early pastoralists probably had a greater impact on forest resources than sawmilling, with extensive stands of timber ringbarked to increase pasture for sheep and cattle. Foster Fyans, commissioner of Crown Lands for the Portland Bay District, reported as early as 1837 on the absence of building timber within a 30-mile radius of Geelong. In 1839 he informed the colonial secretary that squatters were 'disfiguring the country in cutting and destroying the timber, which is not abundant' (in Jones, P. 1981: 230, 235). Restrictions on timber cutting in the 1840s and 1850s had little effect, as demand increased and clearing was regarded as a sign of 'progress' (Wright 1989: 45).

Sawmill licenses were introduced in 1862, the same year as the first reserves were set aside for the preservation and growth of timber. By 1871, however, dwindling supplies and an inability to slow the rate of illegal cutting had prompted the establishment of local forest boards. These were to oversee the management of forest areas, but lack of funds and effective regulations meant the boards achieved little. To replace them, a Central Forest Board was formed in 1874, to supervise all the colony's forests. An 1876 *Forest Act* legislated for local boards again, but these proved to be a failure as well (Wright 1989: 245).

Although these various boards had limited impact on direct forest management, they helped pressure four Bills into parliament (1879, 1881, 1887 and 1892), specifically aimed at forest protection. None were enacted, however, due to entrenched opposition and apathy. Apart from a brief period in the 1870s, timber throughout Victoria was cut under a system of annual licences. This allowed the holder of a miner's right or a carrier's license, the purchaser of Crown land or the licensee of a sheep or cattle run, to cut any quantity or quality of timber without restriction for building, fencing or mining, or for use as industrial or domestic fuel. The only exemptions were crescents of Crown land within ten miles of Melbourne, seven miles of Geelong, one mile from the shore of Port Phillip Bay or within the boundaries of proclaimed townships. Sawmillers and other timber-cutters paid a fee of 25 shillings per quarter (Linge 1979: 339).

The work of John La Gerche, a bailiff-forester in the Creswick State Forest in the 1880s and 1890s, reveals the daily struggle of local officials to reconcile the competing demands of industry, politics and conservation. His work to manage a forest in the public interest was impeded by general government neglect of forestry, lack of resources, mining industry demands, a permissive licensing system and a popular resistance to forest authority (Taylor 1998: 7). While La Gerche fought the depredations of firewood and pit prop cutters in the forests and in the courts, he also promoted silvicultural thinning and plantation development. He worked in a period of slowly shifting attitudes, between the traditional belief in forest clearance as a sign of progress and civilisation, and a growing recognition of the need to conserve forest resources to meet future needs.

Several reports attacked the Victorian government for its failure to provide adequate reservation and supervision of forest areas. F. D.'A. Vincent, for example, of the Indian Forest Service, was invited to inspect and report on the condition of Victoria's forest estate in 1887. He was disappointed by the measures taken to prevent waste of the forests, especially the licensing system. Some years later, Inspector-General B. Ribbentrop, also of the Indian Forest Service, condemned existing laws as outdated, deplored the waste and rampant destruction by fire, and held that revenue from the forest estate was ridiculously small. He attacked the concept that the exploitation of forests for private benefit was the foundation of national wealth, and suggested that the license system, 'which permits a scramble for the forest produce', was the cause of much of the disorder (Ribbentrop 1896: 7).

George Perrin, the Victorian Conservator of Forests, also felt the license system was wasteful and open to abuse. In particular he criticised sleeper hewers who, armed with an axe, a maul, a set of wedges and a five shilling quarterly license, cut all the best young timber which would shortly be fit for sawmilling, depriving nearby mills of enough logs to work, and ruining mill plants worth thousands of pounds (Perrin 1890; Powell 1976: 118-21).

Pressure exerted on the government by these and other scathing reports, by the failed bills and the continuing devastation wrought by bushfires finally resulted in the 1897 Royal Commission on State Forests and Timber Reserves. Directed to inquire into most aspects of forest reservation and management, licence conditions, cutting practices and forestry nurseries, it submitted 14 reports to parliament in the following four years, including detailed studies of the major forest areas of the colony, reports on sleeper hewing, forest royalties, fire protection and water diversion. The fourteenth provided a summary of forestry in other countries, and an assessment of the situation in Victoria.

For most of the nineteenth century, the forests of Victoria had been administered by various offices and branches of the Department of Crown Lands and Survey (Carver 1921: 6). Responsibility for conservation and protection of timber reserves was thus entrusted to a department whose charter was to encourage settlement through alienation of the public estate and clearance of the native vegetation. It was thus only after decades of inquiries, commissions, reports and advice that the government passed the *Forests Act* of 1907. This provided for a State Forests Department under a Minister of Forests, administering permanently reserved forest, controlling forest produce on unoccupied Crown Land, and the collection of royalties. In 1918 an even more comprehensive *Forests Act* was passed, establishing the Forests Commission of Victoria, and a Forestry Fund for the improvement and replanting of state forests. Provided with sufficient funding for the first time, the Commission sought to conserve and develop existing forests on sound forestry principles, establish further plantations of exotic softwoods, and undertake research and experimental work (FCV 1928: 46).

The Great Depression brought misery to many Australians, but also significant benefits to forest resources. Unemployment relief measures included such public works as urban and rural road construction, water supply channels, public buildings and forest improvement works. By the early 1930s numerous work camps were established in Victoria in both native forests and pine plantations. Unemployed men and boys worked in the forests, generally for eight weeks at a time. They were engaged mainly in thinning and ringbarking to encourage the regrowth of young mountain ash and silvertop forests, and the construction of tracks and firebreaks (Moulds 1991: 61-3).

Following the disastrous Black Friday bushfires of 1939, intensive milling operations began almost immediately, to salvage the timber before it rotted. The *Forests Act* of 1939, largely the outcome of the 1939 Royal Commission on Bushfires, further restricted fire lighting in summer months, and directed owners of land within one mile of state forest or national park to destroy flammable material when directed by the Forests Commission. It also required sawmillers to construct dugouts and safety shelters for protection in bushfires, and ensured that all sawmills would henceforth be properly licensed. The *Country Fire Authority Act* of 1944 directed responsibility for fire protection on public lands into the hands of the Forests Commission, and to the Country Fire Authority for most other rural land (Carron 1985: 203).

Over the course of more than 100 years, public administration of Victoria's forest estate had shifted from one of almost complete neglect to one of accepting responsibility for most aspects of management. The waste and loss which characterised much of the nineteenth century was gradually curtailed, as officials grappled with the need to limit damage by bushfires, ensure supplies of timber for present and future needs, and meet other demands on forests such as water supply and recreation. Through it all the forests changed and grew, variously neglected, needed or cherished by all those who depended on the forests and woodlands of Victoria.

Sawmilling technology, 1850-1940

The earliest timber-cutting was slow, laborious and repetitive. It involved falling trees with axes and saws, lopping off the branches, and cross-cutting the trunk into long lengths. These were then hauled to a sawpit and manhandled into position. One man stood above the log and another below, pushing and pulling on a long pit saw to cut the log into timber. The 'underdog' below sometimes wore a bag over his head to prevent the sawdust getting in his eyes. A fair day's work for the two men would produce around 100 superfeet of timber (Griffiths 1992: 43). The first steam-driven sawmills introduced in the 1840s were an immediate success, dramatically increasing productivity. Nevertheless, sawmilling was essentially a transport industry, requiring an efficient means of transfering logs to the mill and timber to market. As only about half the volume of a round log could be converted into square sawn timber, with the rest going to waste, the distance and cost of transporting logs had to be kept to a minimum. Sawmills were thus located close to the supply of timber, as it was generally easier to move a mill to the trees than to haul logs for long distances through rugged forest areas. Mobility was thus a key feature of the industry as well, with mills shifted after a patch of forest land had been cut out. This resulted in a pattern of sawmills distributed throughout forest areas, generally connected by track or tramway to government roads or railways. The average duration of a mill at one location was around five years, although much longer periods at the one site were also common. The longevity of a mill depended on a range of factors, including the forest area available for cutting, distance to market, motive power, workforce and terrain.

Sawmills required a reliable source of water to supply boilers, to lubricate saws and for domestic consumption. A sloping site beside a creek or river was often preferred, to take advantage of gravity in handling logs. Creek valleys and ridges provided workable gradients for the dispatch of timber along tracks or tramlines. Forest sawmills were thus generally located in the midst of their timber supply, close to a water source, and within working distance of road or rail facilities.

Sawmilling remained a labour-intensive industry. It involved cutting down trees, hauling ('snigging') logs, loading, sawing and timber transport. The heavy labour involved in falling trees by axe and crosscut saw meant that logging operations in forest areas were generally based on partial falling, with trees that were diseased, poorly shaped, over-aged or under-sized ignored in favour of suitable mature specimens. After selecting a tree for its size and soundness, holes were cut to put in boards for the axemen to stand on (Plate 3). Then a scarf, or wedge, was cut into the trunk above the buttress roots. The positioning of the scarf was important as it determined the direction of the tree fall, avoided defective timber in the butt, and ensured a more cylindrical log for the mill. The direction the log was to be moved after it had fallen was also considered. A strip of bark was then removed from the opposite side of the tree where the sawcut was to be made, about 15 cm above the bottom of the scarf. As the saw cut into the trunk, wedges was hammered into the incision to prevent the saw from jamming. After the tree had fallen, the crown was removed, and the log was crosscut into shorter lengths of six to seven metres. An efficient crew could fall up to seven or eight trees in a day, with half a dozen men in each crew (Evans 1993: 5).

The removal of logs from the bush involved snigging, or dragging them either to a landing or directly to the mill. Horse and bullock teams were widely used in this role until the Second World War (Plate 4). Bullocks were often preferred, however, because they were cheaper, better at hauling heavy loads, foraged in the bush, and did not require expensive shoeing and harnessing. As early as 1900, however, steam-powered logging winches were becoming increasingly common, especially in forest areas with a dense concentration of timber trees in a limited area (Plate 5). Winches consisted of a boiler and a separate frame, on which was mounted the steam cylinders and cable drums. A cleared path was prepared for the log, which was attached to the 'fishing line', a short wire rope permitting the haulage of logs from the bush to the winch line. Two wire ropes were then attached to the fishing line: the main line, which hauled the log, and a tail rope, to return the main line to the cutting area. The ropes were attached to separate drums on the logging winch, which could each be operated independently (Plate 5). The log was then hauled to the landing, which consisted of a platform of logs laid beside the tramway, and transferred by jack onto a log truck. A horse team then drew the log truck along the tramway into the log yard at the mill. Winch crews generally consisted of a qualified driver, a ropeman, whistle-stringer and woodcutter (McCarthy 1987: 83-4).

Plate 3. Tree Falling (courtesy of the Birregurra District Historical Centre).

Plate 4. A horse team on Hayden's log tramline, Barwon Downs (courtesy of Birregurra District Historical Centre).

Plate 5. Steam-powered logging winch (courtesy of Birregurra District Historical Centre).

Plate 6. Breaking-down bench at Henry's 'Fun Factory' Sawmill (courtesy of Birregurra District Historical Centre).

After the log was brought to the mill, it was cut into large squared slabs (flitches) by the breaking down saw. These saws were freqently twin circular blades, with one mounted above and slightly ahead of the other to allow a clean cut through logs up to six feet in diameter (Plate 6). Flitches were then moved to the rip bench for sawing into the required cross-section. The sawn timber was then cut (docked) to the correct length, and any faulty sections removed. Discarded timber was used as boiler fuel or burnt as waste. Sawdust was removed by shovel and wheelbarrow along a trench dug beneath the main saw bench, and dumped on heaps which grew to immense size if the mill operated at one location for years on end.

The working day for forest sawmill workers generally began at 7.30 a.m. and ended around 5.15 p.m., with an hour's break for lunch. An additional four and a half hours were also worked on Saturday mornings, making a 48-hour week for forest mills. No allowance was made for travelling time, and both mill and bush crews were expected to be in position at starting time. This was especially relevant to bush crews, members of which often worked several miles from mill and settlement. Henry's No.1 Mill consisted of two or three sawyers, several pullers out, the docker, an engine driver, a number of general labourers, a sawdusty, blacksmith and tallyman. The latter was responsible for stacking and calculating the volume of timber on each log truck as it passed from the mill.

The location of many sawmills in remote forest areas meant that tramways were often the only way to transport timber. Existing roads were frequently unsealed, pot-holed and at times impossible to negotiate in winter. An all-weather tramline, however, could be constructed from materials readily at hand, such as wooden sleepers and packing, employing the labour of mill workers or contractors. Timber tramways thus provided the link between logging areas, sawmills and government railways, making it physically and economically feasible to transport timber year-round to metropolitan markets. In addition, tramways provided essential communication between isolated mill settlements and district centres, facilitating the transport of provisions for mill workers and their families, feed for horses, and new parts and tools as needed. Although the tramways used simple technology, the capital cost of their construction could equal that of the sawmill itself, restricting their use to areas of high quality forest. Horses and steam locomotives were the main sources of motive power on tramways until the 1920s. Although horses were cheap, reliable and better suited to hauling on steep grades, larger mills served by longer tramlines often invested in small steam locomotives to convey timber to market (Houghton 1995; Plate 7). After the First World War, modified petrol engine trucks and diesel tractors were also used for tramway haulage (Frawley 1993: 155).

Plate 7. Horse-hauled log tramline (courtesy of Birregurra District Historical Centre).

Tramways were generally characterised by their economy and light construction. Earthworks were minimised by the use of timber bridges spanning gullies, and steam winches were used on steeper grades. After surveying the route, the track bed was cleared of timber, then excavated by pick, shovel and horse-drawn scoop. The tramway itself was constructed of wooden rails nailed onto rough timber sleepers and split packing, commonly at a gauge of three feet. Wooden rails provided the most effective braking surface for the iron tram wheels, which featured deep flanges and treads up to seven inches across. Steel rails were sometimes laid, however, on heavily used sections of track, especially curves and close to the mill, as timber rails wore out very quickly. A tram repairer was often employed to replace rails, repair bridges, and renew packing displaced or broken down by the horses.

Tramway trucks consisted of two small iron and timber bogies. Each truck featured a wooden framed chassis, secured by long transverse iron bolts. Swivelling bolsters were fixed to the top of the chassis, with chains attached to secure the load of logs or timber. Wheel and axle sets on which to mount these were commonly obtained from Melbourne foundries (Houghton 1975: 91). The speed of log trucks was controlled by hauling on ropes connecting the brake levers, which pulled the wooden brake 'blocks' up between the wheels. Each load weighed up to two and a half tons. After arriving at the tramline terminus, loads were transferred to large timber stacks in the railway station yard, prior to dispatch to market.

Victorian sawmills from the 1850s to the 1940s were thus characterised by the use of steam engines and animal motive power, with mills generally owned and operated as small, often family-run businesses. Capital costs for mill and tramline construction were high for such small firms, so that mill plant was commonly acquired secondhand, and sold on again when the operation closed. Large sawmilling companies, such as those that dominated the industry in Western Australia, were rare. Labour requirements were high, which meant that sawmills provided valuable employment opportunities in many forest districts. Local markets, however, were generally small, and competition from cheap imported softwoods continually threatened the economic security of the industry. Eucalypt forest resources offered large but highly variable log sizes, resulting in the establishment of a large number of small sawmills distributed in frequently remote forest locations.

The Otways

The Otway Ranges rise to about 500 metres above sea level, dominating the terrain, climate and vegetation of the region. The steep slopes are the source of numerous creeks, rivers and waterfalls, while the soils of the ranges support a diversity of tall, temperate eucalypt communities.

Prior to the arrival of Europeans in the nineteenth century, Aboriginal people exploited the coastline to the south and the more open country to the north, but their penetration of the forested uplands is less well documented. Pastoralism and timber extraction in the Otways began in the 1830s and 1840s, and selectors began to carve out small farms in the 1860s. The arrival of railways by the end of the nineteenth century set the stage for a dramatic expansion of timber-getting in the region.

Aborigines in the Otways

Several local Aboriginal groups occupied the Otway region. The *Djargurd* lived west of Cape Otway and Pirron Yaloak, and the *Gulidjan* were found around Lake Colac. The *Wadjawuru* lived east of Birregurra, while the forested plateau and the rugged coastline of the Otway peninsula were occupied by the *Gadubanud*. The economy of the coastal groups was based on shellfish collection and fishing in the tidal estuaries, supplemented by plant foods collected from the river flats. The inland forest areas may have been occupied on a seasonal basis by families who spent much of the year either on the coast or in the basaltic plains to the north. Within the forest, koala, possum and wallaby were hunted, the latter with dogs, spears and fire. The relatively low animal populations and difficulty of movement probably restricted utilization of the tall open forests in the wetter parts of the ranges (Clark 1990; Presland 1982; Scarlett 1977; Stuart 1981).

The prehistoric Aboriginal occupation of the Otways is archaeologically little known. Several sites, however, such as a dolerite axe quarry and a camp site at Gellibrand, and a burial cave near Barramunga, indicate that the area was used to exploit raw materials. These included the rare and valuable 'bundit' spears, probably made from the satinwood tree, which were traded for 'maleen' spears from the Mallee (Dawson 1981: 78). It is also possible that remote locations may have been favoured for ceremonial and burial purposes, as well as for collecting tree bark and various food resources. Tracks along ridges and watercourses were probably the preferred means of traversing the ranges.

The impact of European settlement on Aboriginal people in and around the Otways was, as elsewhere, catastrophic. Massacres, introduced diseases and the abduction of Aboriginal women devastated local populations. One group was pursued and murdered in retaliation for the killing of a surveyor in 1846. As Aboriginal society collapsed, a Wesleyan Mission was set up in 1839 at Buntingdale, south of Birregurra, by the Reverend Francis Tuckfield. However, fighting between the disparate groups gathered together there, and the refusal of Aborigines to adopt a settled life, failed to halt the destruction of the Aboriginal people. In 1850, the reserve was opened up for European settlement (McFarlane 1982).

Europeans in the Otways

The first recorded European sighting of the Otway coastline was made in 1800 by Lieutenant James Grant, in command of the brig 'Lady Nelson'. He named Cape Otway after a naval friend, Lieutenant William Albany Otway. Nicolas Baudin surveyed the same waters in 1802, while later in the year Matthew Flinders charted Moonlight Head during his circumnavigation of Australia. The first white men to enter the Otway forests, however, were probably Joseph Gellibrand and George Hesse, who disappeared in 1837 while attempting to trace the Barwon River from Geelong.

Pastoral development of the ranges began in 1839, when William Roadknight secured the Gerangamete run at Barwon Downs. Searching for more grazing land, Roadknight and Thomas Vicary established a route from Colac to Cape Otway. This helped provided access for Governor La Trobe's party in 1846, when, in response to the wrecking of the *Cataraqui* off King Island in 1845, they travelled overland to select a site for a lighthouse overlooking Bass Strait. Further pastoral activity included Captain Fred Dewing's run at Paraparap, established in 1845, 12 miles south of Birregurra on the Barwon River (Billis and Kenyon 1932; Lemon and Morgan 1986).

The earliest timber-cutters in the area began work at Apollo Bay and Lorne in 1849. The first sawmill and jetty were built at Apollo Bay in 1852, supplying railway sleepers, bridge and wharf timbers to Geelong. During the following decade, sawmills in the area provided work for several hundred men and supported a thriving township. Timber removal was limited, however, due to the inaccessibility of the region, and the difficulties of transport. By 1864, shipping masters refused to load timber cargoes at Apollo Bay, citing rough seas, poor loading facilities and the loss of several vessels. Lacking alternative means to transport timber, the mills were abandoned, engines and plant removed, and the tramways demolished. Nevertheless, reports of the quality of the region's timber resources, and the need to relieve excessive demands made upon the forests of central Victoria, encouraged the construction of a new jetty at Apollo Bay in the 1880s, and prompted a minor resurgence in the local industry. Mills were constructed in 1875 on Wild Dog Creek, in 1885 on the Elliot River, and in 1886 on the Barham River (Hebb 1970: 377).

Pressure was also applied in this period by mineral speculators and land-hungry settlers to open the Otway Ranges for development. Land Acts of the 1860s aimed to break up the monopoly of the squatters in Victoria and allow poorer people to select land. The *Land Act* of 1869, for example, permitted selection of up to 320 acres at £1 per acre (or at two shillings per acre for ten years), on the condition that the selector reside on his land, enclose it 'with a good and substantial fence' and cultivate at least one acre out of every ten (Powell 1973: 164-88). By the mid 1870s there were about 100 selectors scattered through the Otway forests. The *Land Act* of 1884 enabled selection of blocks at an annual rental of only twopence per acre, encouraging even more settlers into the Otway forests.

Many had never seen their blocks before arriving in the Otways. Settlers came from Colac and district, from the cities, and from the Wimmera and far west of Victoria. Although many lacked substantial assets or capital, they undertook the immense task of clearing the trees from the rugged landscape on the new farming frontier. Under the lease arrangements of selection, the destruction of scrub and crooked or useless timber was permitted, but this was widely interpreted to mean the ringbarking and burning of all the timber on settlers' blocks. Clearing the land thus meant ringing the trees, which were then felled by axe and crosscut saw, rolled into heaps and burnt. Stumps were either laboriously dragged out by horse or bullock teams, or covered with earth and set alight, to smoulder down to ashes over days or weeks. Alternatively, selectors sometimes made informal arrangements with local sawmillers, who would pay to cut the timber on their land and thereby help to clear it. With a cleared area established, the settler could grow vegetables and other crops, and run a small herd of dairy cows. The lack of roads and railways in the area, however, meant there was little opportunity to market crops or produce. Some settlers built corduroy log roads and sleged their milk and potatoes out during the summer, but in the winter they were confined to their blocks by roads so deep with mud that horses were buried. In the view of one official, the selector 'found magnificent timber on his land, but in order to get artificial grass for a few head of cattle he has had to destroy it' (Report 1899: 6-7).

While selectors slaved to clear their blocks, politicians and bureaucrats struggled to impose competing policy visions on the Otway forests. The Lands Department and the Victorian Railways promoted settlement of the region, to encourage the development of a rural, family farm economy, and to generate greater departmental revenues. They were opposed by the forestry, construction, mining and water interests, all of which opposed further settlement in forest areas, regarding it as a threat to future mineral, timber and water resources. Mining companies feared that opening more forest areas to selection would increase the price of timber and threaten the viability of the mining industry. Successive governments responded to these competing interests with a very haphazard, *ad hoc* process of reserving and excising large acreages of forest in the region.

Throughout the nineteenth century the forests of the Otway Ranges were recognised as one of the most valuable timber resources in Victoria, and among the finest forests in Australia (Ivey 1874: 7; Report 1899: 11). Observers commented not only on the 'grandeur and magnificence' of the trees in the region, but also on their apparent inexhaustibility (Hebb 1970: 387). In 1887 *The Argus* described the Cape Otway forests as 'one great wilderness of timber' (*Argus* 7 Jan. 1887). Members of the 1897 Royal Commission on State Forests and Timber Reserves characterised the range as

> a rugged and broken tract of country traversed by hill ridges, with deep gorges or creek valleys lying between, the whole clothed with a magnificent mixed forest of valuable timber trees (Report 1899: 6).

The flora of the area was noted for its extensive stands of blue gum, mountain grey gum, mountain ash and messmate. The rare and valuable blackwood, myrtle beech and satinwood also grew in the Otway forests, prized for their beautiful grain and suitability for cabinet-making and carriage-building. The great height and girth of trees also filled visitors with astonishment. Specimens of mountain ash were commonly recorded as exceeding 250 feet (76 metres) in height, while a shire engineer measured a fallen trunk at 329 feet to a point where the top had broken off. A hollow tree with a girth of 65 feet at five feet from the ground was used by a selector as a stable for several horses. Numerous specimens of similar size were recorded over the years, while the commercial value of such vast trees was noted as well. The Ditchley Park Hotel at Beech Forest, for example, which comprised 13 rooms, was reported to have been built from the palings and boards of a single mountain ash (Hebb 1970: 385; Report 1899: 5).

Nevertheless, members of the 1897 Royal Commission also lamented the 'simply appalling' destruction of timber in the region. They blamed indiscriminate selection and the slavish removal of trees as the root of the problem, and condemned settlement in the forest as 'a great administrative blunder' (Report 1899: 7). They also criticised the great damage caused by fire, often lit by settlers burning dead scrub and undergrowth, who preferred to wait for days when a hot northerly wind was blowing. Sawmillers claimed that the value of the timber destroyed in the Otways 'would have paid the national debt' (Report 1899: 7). The principal recommendation of the Royal Commission was the expansion of existing reserves in the Otways to protect forest and water resources, and the appointment of a local forest officer to supervise the reserves. Such changes were to take years to implement.

The establishment of rail links in the Otway region provided a significant stimulus to both settlement and sawmilling. A broad gauge line linked Wensleydale to Moriac in 1890, the main purpose of which was to supply boiler fuel for the factories of Geelong. In the following year a branch line opened from Birregurra to Forrest. The economic depression of the 1890s meant that modern sawmilling was slow to be established in the area, but market demand gradually increased and peaked in 1912, before slowing again between 1914 and 1918. Opening of a narrow gauge line between Colac and Beech Forest in 1902, with an extension to Crowes by 1911, encouraged further development in the western Otways, and initiated a boom in timber production during the following years. The new lines prompted sawmillers with an interest in the Ballarat timber market to move into the Otways between 1895 and 1905. So great was the volume of Otway timber transported to the Berringa, Ballarat, Creswick, Rokewood,

Pitfield and Maryborough mines that it was said there was more Otway timber underground than above. The pressure of demand led to calls for direct rail access between the two areas, and resulted in the opening of a rail link in 1911 from Irrewarra via Beeac and Cressy to the Ballarat line (CH 29 Sept. 1911).

Regular supplies of building and fencing material were also despatched to Geelong and the Western District. Splitters sent rails, palings, posts and vine poles to farmers in Koroit, Warrnambool, Port Fairy, Geelong and Ballarat. Selectors frequently paid demurrage to halt trains while posts, palings and laths were loaded from beside the tracks, thus utilising timber on their properties which they were previously obliged to destroy (CH 13 Feb. 1903). Demand for mining timbers in the Central Highlands goldfields, after supplies from the surrounding forests were exhausted, provided an additional boost to the Otways timber industry between 1900 and the First World War. After 1918, the War Services Homes Commission contracted a large portion of its Victorian requirements from the Beech Forest area, resulting in a massive increase in production during the early 1920s, with six mills devoted exclusively to the contract (Houghton 1992: 31).

Extensive sawmilling facilities were established during this period at Forrest, Barongarook, Birregurra, Beech Forest, Barramunga, Barwon Downs, Gellibrand, Lorne and numerous other settlements. Timber tramways provided the essential transport links between forests, sawmills and railways, replacing the bullock teams of the nineteenth century. Bullock tracks in the area were renowned for deteriorating into winter quagmires, sending freight costs soaring and seriously hampering the movement of goods and people. Numerous local roads were officially closed to timber traffic from May to September, owing to the damage which timber drays caused to road surfaces. The road from Forrest to Barramunga was still so bad during winter in the 1920s that the Barramunga Hotel had to obtain its beer supplies before the rain and mud of winter set in (Houghton 1975: 32-3).

The tramway system in the eastern Otways connected to the broad gauge Forrest railway line, which skirted the edge of timber areas and required tramways of up to 35 kilometres in length, greater motive power and sawmills of larger capacity. In the more rugged western Otways, the narrow gauge Crowes line was approached by shorter tramways, rarely longer than ten kilometres, with many more medium and smaller-sized sawmills (Houghton 1988: 296). At Lorne, timber tramlines were constructed in creek and river valleys west of the town, terminating at the jetty on Point Grey. At Wye River, horse-hauled tramways extended into valleys west of the settlement to extract timber for the mill, which by 1919 was one of the largest and most modern in the Otways (Scales 1983). So successful were timber tramways as a mode of transport that more than 300 kilometres of line were laid to the Otway railways (Houghton 1988: 295).

Capital investment in Otways sawmills varied considerably in this period. Family companies of local origin dominated the industry between Colac and Beech Forest, and obtained most of their timber supplies from alienated selections. Along the ridge, from Olangolah to Crowes, access to state forest areas favoured outside companies, although local family businesses operated successfully in the area as well. The large mills at Forrest and Barwon Downs, such as Henry's, Sanderson's and Hayden's, were owned by professional millers moving from cut-out forest areas of central Victoria (Houghton 1988: 294).

Logging methods in the Otway forests, as elsewhere in Victoria, were generally based on partial falling. Sawmillers selected mature trees suitable for their mills, and left standing those that were diseased or poorly shaped. Given the time and labour involved, it was rare for a tree to be wantonly cut and left. Clear-felling only occurred when an even-aged stand of trees was available, and demanded by the market, such as the 1851 bushfire regrowth cut in the 1930s. Alexander Sanderson, a sawmiller based in the Noonday Creek valley, explained to the 1897 Royal Commission that after his crew had 'gone through the forest and taken the timber suitable for the mill there is still a dense forest left' (Minutes of Evidence 1899: 156). His view differed from that of A. C. Ilie, district forester for the east Otways, who in 1925 labelled the system of cutting then in vogue one of 'clearcutting. Every tree of any value is taken', but he acknowledged the practice of leaving overmature, diseased and dead trees standing (Ilie 1925: 8). Splitters also plied their trade, cutting trees in deep gullies that were otherwise impossible to exploit. They received about four shillings per 100 palings from the sawmiller in whose patch of forest they worked. The split timber was then carted out of the forest on tramlines to market.

Minor forest industries also emerged in the Otway region in the early part of the twentieth century. Charcoal burning commenced at Barongarook in 1916, using a circular iron kiln, while wattle-barking was also practised at the settlement. Wattle bark was valued for its tannin by the leather tanning industry, and bakers preferred the burning and heating qualities of wattle wood, so that trees were cut down and the wood sold after the bark was removed. The forests around Barongarook also supplied great quantities of firewood for industrial and domestic use at Colac and surrounding settlements, used in steam boilers, wood-fired stoves and open hearth fireplaces (Houghton 1979: 31-2). Eucalyptus oil was distilled during the early 1920s on Ten Mile Creek, while the first sawdust-burning brick kiln in the Forrest area was built in 1934 at Sharp's Tanybryn mill, south of Barramunga (Houghton 1976: 17).

Following the conclusion of the First World War in 1918, a large stock of ex-Army vehicles came onto the market, and road transport began to compete with rail delivery. The 1920s and 1930s saw the gradual replacement in the Otways of steam boilers and animal-powered log

Plate 8. Early log truck (courtesy of Birregurra District Historical Centre).

transport by the internal combustion engine (Plate 8). The Country Roads Board and the Otway Shire Council (formed in 1919), upgraded bullock tracks and roads, and motor trucks began carrying more of the region's timber output. Rail-tractors also became an increasingly common form of tramway haulage. The introduction of bulldozers with winches in the late 1930s provided greater mobility than fixed tramlines, and access to formerly unworkable patches of forest, especially south of Beech Forest and Barramunga. Following the devastation of the 1919, 1926 and 1939 wildfires, and the disruptions of the Second World War, timber tramways became increasingly outmoded, and gradually deteriorated through age, fire and neglect.

After 1939, the Forests Commission encouraged the centralisation of sawmills into towns, with mill workers and their families moving out of the forest as well. In the Otway region, large mills were established at Forrest, Deans Marsh, Gellibrand and Colac. The Commission also enforced stricter regulations on the use of such fire hazards as steam winches and locomotives. This meant that sawmills were no longer restricted in supply to a single logging area, previously granted to the miller as a specific acreage or radius around the mill. Instead, logs could be selected for each mill from several sites by the use of large motor trucks, tractors and bulldozers. The last timber tramway to work in the Otway forests, at Hayden Brothers' mill at Barwon Downs, closed in 1972 (Houghton 1975: 91-2).

Henry's No.1 Mill

Thomas Cowley, a timber merchant from Ballarat, was one of the first sawmillers to exploit the opportunities provided by completion of the Birregurra-Forrest railway line in 1891. In that year he established a mill at an isolated location near the top of the Otway divide. The mill was managed by Will Rodgers until 1896, when Hayden Brothers of Barwon Downs took over the lease for 18 months. Cowley shifted his operations to the valley of the Noonday Creek in 1897, establishing a new sawmill 6.5 kilometres south of Forrest. The site was located at the junction of the Monday and Noonday Creeks, and was connected to the Forrest railway station by a horse-hauled iron tramway. The Sanderson family also established a sawmill in 1897 in the Noonday Creek, but closer to Forrest, resulting in a parallel set of tramlines extending up the valley, and disputes over crossing points at the Barwon-Noonday junction (Houghton 1975: 51).

Cowley's Noonday mill was of a standard size for this period, cutting about 7000 superfeet (*c.*16.5 cubic metres) per day, and was surrounded by a small settlement of huts and houses. More than 30 men were employed. The business traded as 'Otway Sawmills, Forrest', and in 1901 was expanded to include James Whitelaw and John McGregor. The main log tramline feeding this mill extended three kilometres up the Noonday valley to the south-east, before terminating at a landing on the tip of a shallow spur. Logs in this lease area were up to five feet in diameter, and hauled down the slopes by teams of animals. Beyond the landing the creek valley narrowed to a steep-sided gorge, preventing further extension of the tramline. Crowley's logging crews could only cut timber from the high ground, and left the side gullies untouched. The gorge areas further upstream also restricted cutting, resulting in the mill soon running out of accessible timber (Houghton forthcoming: 107-8).

The firm applied successfully for a new forest lease to the south, in the watershed of the Barwon River. In 1885 the area had been made a temporary water reserve for the Geelong water supply. In 1899 the area was recognised by members of the Royal Commission on State Forests and Timber Reserves as one of several 'extensive tracts of valuable timber land' excluded from reservation in the Otways State Forest (Report 1899:1). Edwin Robbins, a sawmiller in the Barramunga area, testified to the Commission on the extent and quality of the timber resources in the West Barwon area:

> I have been over a good bit of the Barwon water reserve. There is a tremendous lot of timber on it. I think it should be proclaimed a State forest and retained as a permanent reserve [i.e. for timber production] (Minutes of Evidence 23 Feb. 1899).

The proprietors of Otway Sawmills also recognised that the vast, untouched timber reserves in the West Barwon valley to the south guaranteed a secure timber supply for up to 20 years. Gaining access to the forest, however, was difficult. One possibility was to abandon the Noonday Creek tramline and construct 8.5 kilometres of new line to the mill site. Not only was the cost of such an option prohibitive, but the Sanderson family had relocated their sawmill to the Barwon valley in 1900, 6.5 kilometres south of Forrest, and refused permission for Otway Sawmills to build a tramline through the timber lease which they, the Sandersons, held. The only alternative open to Cowley and his partners was to cross the spur which separated the Noonday Creek from the West Barwon valley, extending their existing tramline either through a tunnel or via a winch-powered incline over the narrowest part of the spur. An incline would have required the ongoing expense of a permanent steam winch with driver and woodcutter. Recognition of this prompted the building of the tunnel, for which W. R. Henry was invited to join the partnership in 1902 (Houghton 1975: 53).

Although a plumber by trade, Henry had achieved success in goldmining at Kalgoorlie and Ballarat. His involvement in Otway Sawmills and the proposed tramway tunnel stemmed from his experience in mine management, his willingness to obtain the necessary finance, and an undertaking to buy out his partners' interests following its completion. It was a risky venture. Not only was the tram tunnel an unusual and expensive option, but construction of the Lilydale-Warburton railway line between 1899 and 1901 was bringing the high-value forests of the Upper Yarra district into commercial production, in direct competition with Otways timber for the Melbourne market. In addition, a slow recovery in the local industry from the 1890s depression was further restricted by the removal of tariffs on timber imported into Victoria following Federation, bringing quantities of Tasmanian, American, Norwegian and New Zealand timber onto the market as well (Linge 1979: 343-4).

Construction of the tunnel initially involved re-laying a former logging tram route with iron rails, and extending it 300 metres further up the Noonday Creek valley (Plate 2). Ballarat miners were hired to dig the 950 feet (290 metre) long tunnel, which emerged high on a spur overlooking the Fork Paddocks and West Barwon River. The tunnel was eight feet high and sloped from eight feet wide on the floor to seven feet six inches at the roof. It was 'timbered throughout with props, cap pieces and laths, like a main drive in a mine', with an estimated cost of a little over £1000 (CH 22 July 1902). The first section of the tunnel was dug with a slight curve to permit the opening to avoid a stream of water flowing down from a higher level (Plate 9). The rest of the tunnel was straight but the bend made it quite dark inside (Houghton 1975: 53). The tramline was laid through the tunnel in October 1903, and extended 1.5 kilometres up the valley to the site of the new mill. The line appears to have been laid with iron rails. In this period the prospect of a tramline into an uninhabited area which lacked roads was a developmental asset for which government financial assistance could be made available. The Victorian Railways (VR) thus leased or sold rails on the principal that they could not receive any timber freight if a tramline did not exist. Thomas Cowley, Alexander Sanderson and Hayden Brothers obtained their rails in this fashion (Houghton 1975: 33). The marking of contemporary parish plans with 'Victorian Railways Commissioners' at the site of Henry's Mill suggests that rails to the new mill were also acquired under VR provisions.

The mill was established in 1904 on the only wide and relatively level area in the river valley south of the Fork Paddocks. Henry was recognised as the dominant partner in the firm, and began to buy out his partners' interests. Thomas Cowley was the first to leave 'Otway Sawmills', departing in 1903, and by July of that year the partnership was known as 'Whitelaw, Henry & McGregor'. Along with sawmilling interests in the Noonday and Barwon valleys, the firm were also timber merchants, with a business in Ballarat (Houghton 1975: 56; Sands and McDougall 1904). In October 1904, the partnership was reduced to 'Henry & McGregor, but it was not until 1907 that Henry began trading as 'Henry & Son', with Henry's son Ernest (born 1885) joining the business (Sands and McDougall 1907). John McGregor subsequently maintained an involvement in the Kalimna sawmill at Lorne until 1911.

Construction of housing at the mill was under way by early in 1904. Wooden cottages for workers with resident families were built some distance to the north-west of

Plate 9. The northern entrance to Henry's main tram tunnel (courtesy of Birregurra District Historical Centre).

the mill shed, in an area known as 'Port Arthur'. The origins of this nickname are obscure. It may relate to the nineteenth-century Tasmanian penal establishment, emphasising the isolation, hard work and basic amenities of each environment. The name may also play on the presence of Tasmanian workers at the mill, although this remains undocumented. Alternatively, the name may have derived from the harbour of Port Arthur on the Liaotung Peninsula in southern Manchuria. Seized by the Russians in 1898, who hoped it would provide a new terminus and warm water harbour for the Trans-Siberian railway, Port Arthur was besieged and eventually defeated by the Japanese in 1904-05 (Tupper 1965: 319-32). The decisive battles of the war were fought at the same time as the mill was under construction. Perhaps the forest workers empathised with those Russian troops trapped in a remote and forbidding place, at the end of a distant railway line. The origins and grim humour of the name is reinforced by that given later to Henry's No.2 Mill, higher and even further isolated than the No.1 Mill. It was known simply as 'Siberia'.

The sawmill shed dominated the No.1 Mill. It was the largest building on site and the centre of activity, with tramlines converging upon it from the west, east and south. In construction it was essentially a large corrugated iron roof supported by piles, open on three sides to permit the easier handling of logs and timber. When the largest logs were rolled off tram trucks onto the landing beside the shed, the whole mill shook. Inside was a whirring, clanking array of belts, benches and saws, driven by a double four-cylinder Thompson engine built in Castlemaine, with pressure provided by a large Cowley boiler. The mill crew of 25 or more included sawyers, an engine driver, labourers, a tallyman, sawdusty and foreman. The shed lay within a constant fug of boiler steam and smoke drifting up from the smouldering sawdust pile, and from the 'heart heap' nearby which burned constantly to dispose of off-cuts and logs with a rotten core. There is no evidence, however, that a dugout was ever constructed for the safety of workers in the event of bushfire (Plate 10). Logs for the No.1 Mill were initially obtained from the hill slopes between the tram tunnel and the mill. When this area was cut out the log tram was extended east along the valley, laid close to the river across a succession of timber bridges. Fifteen kilometres of tramline were built in this direction to tap the headwaters of the West Barwon River. All were planned and constructed by Cecil Turner and George Hall, two of Henry's long-term employees. Branch lines were built up side gullies, with logs pulled down by animal teams or steam-powered winches. These branch lines employed a mixture of iron rails on the bridges and curves and timber rails along the remainder. The log line near the Coldwater Creek junction passed through a short, second tunnel, 11 metres in length, excavated through a projecting spur on the river. East of this tunnel the tram route extended higher up the valley, surmounted a small waterfall, and then branched in three directions. One log line proceeded up the north slope, while the main line continued along the river and branched up the valley in steep zigzags to gain elevation. A third log line was built south-eastward, to the site of Henry's No.2 Mill (Houghton 1975: 56-60).

Plate 10. View of Henry's No. 1 Mill, facing south (courtesy of Birregurra District Historical Centre).

Plate 11. Henry's Hunslet locomotive (courtesy of Birregurra District Historical Centre).

Horses were used on the tramline between the No.1 Mill and Forrest until 1911, when Henry purchased a small Beyer Peacock locomotive from the Wallaroo and Moonta Mining and Smelting Company in South Australia. Known locally as 'Tom Thumb', it soon became evident that the locomotive was underpowered for the hauling required, so it was restricted to running the level 3.5 kilometres of river line between the main tunnel, past the mill shed, and on to the three-line interchange and the route to the No.2 Mill. The engine took out empty trucks and returned with sawn timber and logs brought to that point by horse teams. Henry then ordered a brand new, more powerful locomotive from the Hunslet Engine Company of Leeds (Plate 11). The machine was despatched in October 1912, and arrived at Forrest several months later. Nicknamed 'the Little Green Beetle' for its olive green colour, the engine was based at the Forrest railway station with a shed and water tank for servicing. It featured a hinged tunnel, allowing it to pass safely through the tunnel on its daily return trip to the No.1 Mill, taking out empty trucks, supplies and passengers, and returning with sawn timber to the railhead (Houghton 1995).

Although the Henrys began using steam locomotives for timber hauling in 1911, horse teams were retained for those steeper parts of the tram network higher up the valley. Good quality horses were needed for the sawmilling industry because the 'rough and tumble character' of bush work meant they had a shorter working life (CH 22 July 1902). Drivers of timber trucks were also responsible for feeding and grooming the horses, with large stables at the mill built close to the single men's huts. A dam on the river behind the mill was deliberately formed where the sawdust pile sloped down to provide a swimming hole for the horse teams.

The mill also employed a full-time blacksmith. From 1921 until 1927, Frederick (Alec) Culph worked his forge with its large bellows adjacent to the mill shed. He kept a white kitten, blackened with soot, which would ride on the bellows and keep warm, close to the fire. The smith was responsible for shoeing all the mill's working horses, as well as making and repairing the iron components of the timber trucks, including shackle and brake block pins, and swinging bolsters. Shoes for each horse hung from pegs on a rail on the wall. The blacksmith also sharpened the saw blades, with a spare set stored in clamps, available whenever those in use became blunt. In addition, he met the numerous other demands for tools and metal equipment around the mill and settlement (R. Henry 12 June 1999).

The years prior to the First World War saw increasing demand for Otways timber. Henry & McGregor established their record in 1905 when they despatched 90 trucks from the Forrest railway station in March alone, while Alexander Sanderson sent 490 trucks of timber to Ballarat in 1906 (CH 7 April 1905, 6 Feb. 1907). Henry purchased an interest in the Geelong timber yard of Thomas Higgans in 1909, to provide a marketing centre for his firm's business. The yard traded under the Higgans name until 1919. Geelong was the main market for Otways timber, although large orders were also sent to Melbourne, Ballarat and the Western District.

W. R. Henry and Son installed their No.2 Mill in 1911, located about four kilometres south-east of the No.1 Mill, close to the top of the Otway divide (Plate 12). About a dozen employees lived at the site. Logs obtained from the tramline west of the mill were probably sent to the main

Plate 12. Henry's No. 2 Mill, c. 1912 (courtesy of Birregurra District Historical Centre).

mill, as the steep gradient made hauling log trucks up to the No.2 Mill very difficult. A log tram was later laid east of the mill, following an easy contour around the slope of a hill (Houghton forthcoming:122). On 15 February 1919, the No.2 Mill, tramline and all its bridging were destroyed by bushfire, and the mill site was abandoned (*Argus* 20 Feb. 1919). Several weeks later, floods caused further damage to Henry's tramlines, washing away a large number of bridges in the West Barwon valley (CH 10 March 1919).

A new mill was established 3.5 kilometres south-west of the No.2 site, near the top of the Coldwater Creek catchment, in an extensive stand of fire-damaged timber. It was known as the 'Buffalo' mill because it was powered by a Buffalo Pitts traction engine that had been driven to the site from Forrest under its own steam. Access to this mill was gained via an incline and winch over the spur south of the No.1 Mill, and along a tramline extending more than five kilometres to the south-east. It operated until 1925 (Houghton 1975: 66-9).

Henry was also engaged in a dispute with the Forests Commission in 1921 over royalty rates. It was a policy of the FCV not to discriminate against sawmillers working in more remote or poorer quality forests, using variable royalty rates to equalise such disadvantages. The Commission set out rates on three classes of timber: A, B and C. Class A was the highest rate, payable on timber of high quality from easily accessible forest, while Class C was the lowest rate, payed on timber of lower quality transported from a greater distance. In February 1921, the Commission reduced the royalty for east Otway sawmillers to 1s 1d per 100 feet, but amended this decision in April, bringing the royalty back to 1s 3d, equivalent to Class A and in line with 'other millers operating in timber of good quality' (FCV 29/0474: 15 April 1921). Henry complained that the revised rate was too high, suggesting that the poorer quality timber available in his logging areas was ground for special consideration. He argued that the Warburton sawmillers had a commercial advantage, with easier forests to log, and cheaper freight costs to major metropolitan markets, and could thus afford to pay the higher rate. He also complained of the loss of his Ballarat customers to Melbourne sawmillers. His final defence was that the £2000 worth of plant fixed at his No.1 Mill forced him to continue operating, otherwise 'we would not stay here a week' (FCV 29/0474: 25 April 1921).

The Forests Commission responded that the east Otway millers had the advantage of the main Western District markets from Geelong to Port Fairy and Hamilton, and they did not have to bear the break of gauge of costs incurred by those in the Beech Forest district, who were subject to handling charges when their timber was transferred from the 2'6" narrow gauge line at Colac. The Commission also felt that the regular demand for hardwood timbers, and rising timber prices associated with the post-World War I building boom justified the increased rate, and the decision remained in place (FCV 29/0474: 23 May 1921).

Torrential rain in the winter of 1923 brought floods down the West Barwon River, washing away 6.5 kilometres of Henry's tramline on 4 June, including about 1500 metres of bridges and makeups, along with several huts from the main mill (FCV 29/0474: 4 August 1923). Most of the river flats and banks along which the line had run were scoured away by the flood. Two winches were also left stranded in the forest, and log supply to the No.1 Mill was abruptly terminated. In the five months it took to repair the damage Henry kept about 20 families and a few single men on the payroll, at a cost of about £100 per week. During this period he had the men cleaning up the wreckage, salvaging and straightening rails from the river, and making temporary tracks to recover the winches and other plant remaining in the forest (FCV 29/0474: 10 Aug. 1923; Davies 1999: 251). Henry used the flood destruction to argue for a 3d reduction in royalty to 6d per 100 feet, down from the pre-flood rate of 1s 3d. It was not until September that the Forests Commission conceded to a fixed royalty of 7d for a period of two years (FCV 29/0474: 4 Sept. 1923).

W. R. Henry & Son may not, however, have been in such dire financial straits as their correspondence suggests. Since 1920, the firm had recommenced working the Noonday mill, cutting mostly messmate and mountain grey gum from areas unlogged 20 years previously. Following the 1923 floods, the mill and settlement were fully re-established. The No.3 (Buffalo) Mill also continued working throughout this period (FCV 29/0474: 18 June 1923). Henry was by now, at the age of 65, an old hand at weathering the market fluctuations for such commodities as gold and timber. He knew that four million superfeet of timber remained to be cut on the ridges east of the No.1 Mill, timber which markets still demanded, if only he could reconstruct his operations in reasonable time. The No.1 Mill was eventually re-opened in November 1923, and the damaged tramline reconstructed several months later (FCV 29/0474: 3 Nov. 1923).

In February 1925 Henry applied successfully for permission to move the No.3 Mill about one kilometre southward over the Otway divide into the headwaters of Carisbrook Creek. The District Forester supported the proposed move, citing the availability of 'Timber first quality', with ten milling trees to the acre, consisting of messmate and mountain grey gum on the ridges, and mountain ash, spotted gum, blackwood and beech in the gullies. He also pointed out how extended Henry's transport lines had become, involving 12 miles (19 kilometres) of haulage to the railhead at Forrest, with horse-drawn log trams and a hauling winch connecting to the steel tramline at the No.1 Mill. Implicit in the forester's recommendations to support Henry's application of a 900 acre lease was concern that his was the only operation likely to utilise the timber in such a remote area, thereby generating revenue and employment (FCV 29/0474: 3 Feb. 1925).

In the same period, poor quality forest trees hampered logging operations for the No.1 Mill. Four kilometres of tramline were constructed after the winter floods of 1923, at a cost of £1100, through difficult country to tap supplies on top of the Otway divide. Trees which appeared outwardly sound were so shattered upon falling that frequently not a millable log was obtained. The Forests Commission agreed to readjust the royalty to 9d per 100 feet for the No.1 Mill, but increased the rate for the No.4 Mill to 1s 3d, given the high quality forest of mixed species available for cutting (FCV 29/0474: 9 Jan. 1926; 8 Feb. 1926).

In the early morning hours of Wednesday, 26 October 1927, a spark from a smouldering rubbish heap blew under the breaking down saws at the No.1 Mill, igniting a fire which burnt the mill shed to the ground, destroying belting, saws and other equipment, and scorching nearby buildings. Timber stacked nearby, however, was largely untouched. The building and plant were not covered by insurance. At the time of the mill's destruction, demand for timber was high, with numerous orders on hand for fulfilment. Given the faulty, over-mature timber available on the Upper Barwon ridge, the decision was made not to re-establish the No.1 Mill, but to consolidate operations at the Noonday site (CH 31 Oct. 1927).

The No.4 Mill was closed in October 1928, due to the long haulage involved in transporting logs and sawn timber. The Noonday mill continued cutting out 'a little bit of back country', with the last logs for the mill winched from the tunnel early in 1929. The mill continued working until March 1929, after which everything salvageable from the Noonday valley was removed, including boilers, winches, huts, houses and rails. The tunnel portals were blown in, and the sound of timber-getting ceased in the Barwon valley for the first time since 1897 (FCV 29/0474: 25 Feb. 1929).

Falling demand for timber in the late 1920s, the Timber Strike of 1929, and the onset of the Depression combined to substantially curtail timber production throughout Victoria. W. R. Henry & Son shifted their activity closer to Forrest and worked only a single mill, in contrast to previous years when two or three sawmills worked in company. Henry assumed management of a mill west of Forrest in 1929, set up two years earlier by Bernard Seebeck, former head sawyer at the No.1 Mill. This area was cut out in 1932, and another mill began operating by Roadknight Creek, about one kilometre to the south-west (Houghton 1975: 69-71).

W. R. Henry retired from active management of the business in 1934, at the age of 76, handing full control over to his son, Ernest. Around the same time, the firm commenced shipping large quantities of Otways timber to Adelaide, running a ketch, the *Argosy Lemahl*, full-time between Geelong and the South Australian capital. In 1935 the firm's main timber merchant business in Geelong was destroyed by fire. New premises were installed which featured a much larger yard, saws, a moulding mill and seasoning kilns (GA 26 Nov. 1957). The firm also began construction work on a tramway to tap a large new timber lease to the south of Mount Sabine. Although the lease had been granted in 1930, the sawmilling industry had not been active enough to justify the heavy expenditure needed to exploit the area. The new tramline commenced beside the Forrest-Apollo Bay Road, 14 kilometres south of Forrest, with a gantry installed to transfer timber to road vehicles. A mill, known as the 'Fun Factory', was established along the tramline, and worked from 1935 to 1937, maintaining production while more mills were erected (Houghton 1975: 71).

As the Fun Factory mill closed a new mill opened near Nettle Creek, drawing logs from steep slopes to the south. The Carisbrook Creek mill commenced working in 1938, and continued in operation until 1948, when the firm built a new mill in Forrest. The Nettle mill worked until 1943, when wartime labour shortages and a diminishing log supply forced its closure. Post-war years saw a new sawmill built in the Forrest railway yard, using equipment and fittings salvaged from the Nettle and Carisbrook mills. This mill was operated by the Henrys until 1972, while the Geelong timber yard remained in the family name until 1993 (Houghton forthcoming: 146).

W. R. Henry & Son were thus a prominent sawmilling firm in the Otway region, active for around 70 years. Their sawmills and tramlines opened up previously inaccessible areas, and provided employment for several generations of workers. They bore the vicissitudes of fire, flood and economic trauma successfully, adapting conventional technology to the challenges of timber-getting in remote, mountain forest environments. Over the years their mills produced large volumes of timber, generating revenue for government, railways and surrounding townships. At the same time, these mills and forest settlements were home for scores of workers and their families. Building houses, gardens and other amenities, they created communities deep in the forest, on the periphery of mainstream social and economic life.

Chapter 3

Excavation Results

In this Chapter, I present the archaeological data which form the basis for analysis and interpretation of issues in later chapters, including settlement structure, consumption practices and health and medicine. The Chapter begins with the methods and results of field survey, followed by descriptions of each area of excavation. Finds are discussed in detail in Chapter 4. The chapter concludes with a consideration of site abandonment processes.

Initial assessment of the No.1 Mill site revealed that it lay in a large, overgrown forest clearing, and extended for a considerable distance along the south bank of the West Barwon River (Figure 4; Plate 1). The industrial part of the site yielded evidence for discarded bricks and tram wheels, while in the residential area to the north-west lay scattered whole and broken bottles, iron piping, occasional ceramics, and low mounds of mixed clay, stone and brick rubble. Linking these two areas of the site was the earth cutting for a timber tramway. The mill's location at the bottom of a steep valley in the Otways State Forest had ensured minimal post-1929 disturbance by fire and logging. Many other sawmills of a similar scale and period of occupation, however, show clear signs of interference due to fossickers, vehicle track construction, land clearance, forestry operations and bushfire. Henry's No.1 Mill was thus selected for detailed investigation on the basis of its accessibility, integrity (i.e. minimal post-abandonment disturbance) and potential for archaeological recovery of artefacts and cultural features.

Site Survey

The survey strategy for the No.1 Mill was determined largely by the site's thick vegetation cover, which hindered movement and surface visibility. Formal survey of the site began with a tape, compass and clinometer traverse of the tramline cutting, which extended in a long, curved 400-metre route through the site. Over this distance it descended 7.80 metres from east to west, or less than one degree (49'). This tramline survey provided a baseline for subsequent site mapping. A similar traverse was then conducted of the West Barwon River, to provide a geographical basis for site planning. The industrial part of Henry's Mill extended 150 metres east-west along the river, and 140 metres back from it to the foot of a spur. The residential area to the north-west occupied a gently sloping terrace above the river, extending 100 metres east-west. The river cutting winding past the site was generally 12-15 metres across, with low summer water levels flowing over a bed of river cobbles.

No historical records of the plan or scale of the mill and settlement are preserved. A basic site plan prepared by Houghton from photographs, oral testimony and physical inspection, however, provided a useful guide to planning fieldwork (Houghton 1975: 57; 1995: 9). Intensive survey of the site identified surface artefacts, cultural features and architectural remains. Attention focused on two large areas of the site which had been dominated by residential buildings: 'Port Arthur', with its family houses, and the south-east section, with single men's huts, married men's quarters and W. R. Henry's cottage. Systematic survey of the latter area involved establishing an 80-metre baseline from the tram cutting to the southernmost extremity of the site, from which right angle traverses were taken every three metres, extending to the gullies entering the site from the south-east and south-west. The visibility of cultural features in this part of the site was severely reduced due to the dense growth of endemic and opportunistic weeds and shrubs.

Results of survey in the south-eastern part of the site provided minimal indication of surface material, with only seven small clusters of artefacts identified. These consisted of either discarded bottles or rusted lengths of iron, lacking any association with architectural remains. No traces of buildings were identified on the surface. This result suggested either that less substantial structures and less intensive build-up of material occurred in this part of the site compared to other areas, or that it was subject to more detailed clearance upon abandonment of the mill. Only in one small area of the single men's huts, where several bottles and a discarded tram wheel were found together, was excavation regarded as justified, to obtain a sample of material for comparison with other parts of the site. Trench D was thus established here.

Survey transects to the north-west provided much greater evidence of cultural material. Surface visibility was higher in this part of the site due to the widespread growth of sheltering tree ferns (*Dicksonia antarctica*), and the lighter growth of smaller shrubs and herbs. During the survey, a number of mature, exotic trees were identified, including two Scots pines (*P. silvestris*), three oaks (*Quercus sp.*), an overgrown South American *Cestrum elegans*, and a large sprawling *Hydrangea*, 17 metres across and up to three metres in height. Eight clusters of artefacts were identified, appearing to represent areas of former activity, discard and abandonment, distributed over an area 100 metres in extent. For the most part, each of these was concentrated in a small area, two to four metres across, and characterised by stone, brick and clay rubble. Each appeared to represent hearth foundations of individual house sites. Their number and spatial distribution corresponded well with photographic evidence and Houghton's plan of the mill (1995: 9). Artefacts associated with each cluster of material were most commonly whole and broken bottles, cast iron oven components, ceramic sherds and rusted lengths of iron rail and piping. On the basis of this survey, three areas were selected for excavation, and established as Trenches A, B and E. Quantities of bottle glass exposed on the surface adjacent to the tramline nearby were collected as Trench C.

Excavation strategy and recording

Excavation took place between 15 January and 1 February 1998. Assistance was provided by students and staff from La Trobe University and Melbourne University. Excavation trenches were gridded on a north-south alignment over prominent surface features. These trenches were then spatially plotted in relation to the site survey baseline. Loose surface artefacts were collected for later analysis. Larger artefacts which were too heavy to remove from the site, such as iron rails and stove components, were recorded by measuring, drawing, photographing, weighing and cataloguing each piece. Trenches were then cleaned of surface leaf litter before excavation proper commenced.

A total area of 70.5 m^2 was exposed in excavation of Trenches A, B and E. This constituted 0.91% of the surface area of 'Port Arthur', and 0.31% of the entire site. Each of these trenches represented a separate house site. The maximum depth of cultural deposit was 20 centimetres. No pit features were identified. Trench C was not excavated, but subject to intensive surface collection. Trench D was a one metre square sounding in the area of the single men's huts. No excavation in the industrial part of the mill was undertaken.

Deposits were excavated by hand using the Harris matrix system. Material was sieved in five millimetre screens and artefacts were bulk-bagged by fabric and recorded by context. Context numbers are highlighted below in bold. Stratigraphic matrices of contexts were also established for each trench. Special finds were those artefacts recovered with potential for dating, including coins, whole bottles and ceramics with makers' marks. These finds were recorded three dimensionally, and stored separately from other artefacts to protect them from handling damage and to aid in their efficient processing. Twenty special finds were recovered during the course of excavation.

Trench A

The surface of Trench A was dominated by a large, low mound of clay up to five metres across, from which numerous bricks and stones had eroded. Surface artefacts included bottles, ceramic sherds, iron piping and short lengths of rail. There were also numerous components of a cast iron kitchen stove, including the front and top plates, fire box base, hot plate lids and flue base. A 5 m x 5 m trench was established under the canopy of tree ferns. These features appear to represent the remains of an individual house site. Excavation data are given below, with more detailed analysis and interpretation provided in Chapter 5 (Spaces and Structures).

Two separate hearth features were identified under the clay mound of Trench A (Figure 5; Plate 13). The northern feature (context **8**) consisted of a clay platform, 1.50 metres square, with rounded, possibly eroded corners. Charred wood on the south and east sides suggest that the platform was originally framed with timber. Above this platform was built a small square brick structure, a single course in height, measuring 80 cm by 70 cm across (Plate 14). This brick structure sloped down slightly to the north-east. Abutting its western edge was an eroded deposit of very hard yellow clay with orange and charcoal inclusions. The east and west sides were formed by three bricks laid side on and end to end, between which lay a flat base of whole and half bricks. The central part of this base was burnt and friable, with a black charcoal stain on several vitrified bricks. Resting above and parallel with the western edge of this brick structure was a short length of iron rail, 70 cm long. Several metres to the south-east lay a similar length of rail on the surface, apparently removed from its original position. This clay and brick feature (context **8**) appears to have functioned as a simple open hearth, with the rails forming a makeshift grate. It remains uncertain from which direction the hearth was originally approached, although the possible remains of a clay backing or low insulating wall on the north-west side suggests likely access from the south-east.

Immediately south of this hearth structure, and separated from it by a narrow, 30 cm-wide channel, was a similar feature (context **9**). It also consisted of a clay platform, measuring 1.40 metres by 1.10 metres, on top of which was built a simple brick emplacement. This structure formed a U shape, measuring 80 cm by 55 cm, open to the south-east, with nine bricks laid flat and end to end. It appears most likely that these bricks formed a base for a kitchen stove, similar in construction to the oven foundations uncovered in Trenches B and E. The cast iron stove components identified on the surface correspond in size with and appear to derive from this hearth feature. Around the north, south and west sides of the oven a substantial clay, brick and stone packing was erected for insulation and protection.

Some indication of the building within which the hearth features were enclosed can be gained from associated burnt deposits. A substantial burnt beam (context **22**), 93 cm long, 12 cm wide and 20 mm thick was exposed, abutting the south side of hearth feature **9** and extending to the south-east. Along with nails and more wood charcoal,

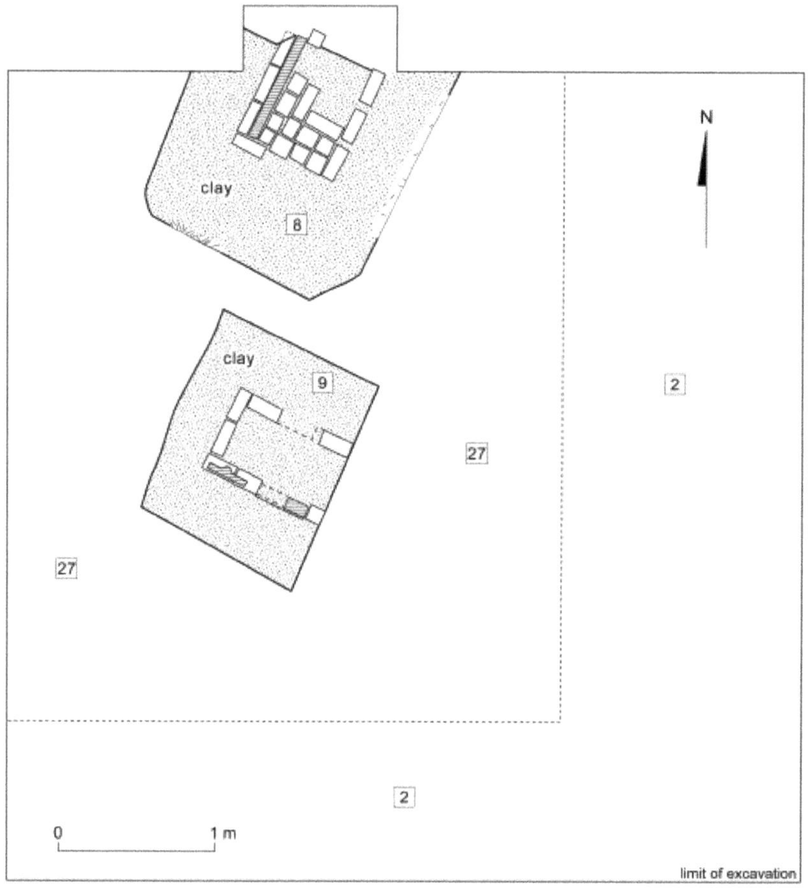

Figure 5. Plan of House A.

Plate 13. House A, Henry's Mill: General view of excavated features (photo: author).

Plate 14. House A, Henry's Mill: Northern hearth feature (photo: author).

this material appears to have formed part of the original timber floor construction. Fragmentary corrugated iron to the north-east of **9** may represent remains of the roof or chimney lining. If the clay and brick feature **8** is interpreted as an open hearth, and feature **9** as a kitchen oven foundation, it is likely that each would have been built in separate functional areas. There are too few remains of the architecture of the building, however, to indicate securely the nature of such areas.

Hearth features **8** and **9** appear to be contemporaneous. Each was built over an underlying mixed deposit of cultural material (contexts **21**, **31**, and **39**), which yielded substantial quantities of broken glass, ceramics and iron. This deposit in turn overlay the natural clay (context **27**). It may have resulted from general domestic discard in the area prior to the construction of the buildings, or have been used as a levelling deposit to fill an uneven surface. A 1913 halfpenny recovered from the deposit stratigraphically below the hearth features (context **21**) supports their post-1913 construction.

Abandonment and collapse of this building was indicated by a broad charcoal deposit, copious quantities of nails, and widespread domestic debris (contexts **3** and **5**). The substantial clay, stone and brick packing (context **4**) around the sides of the hearth (context **9**) weathered and slumped following abandonment of the building. The rarity of bricks in such construction contexts is suggested by the ratio of nine large stones (*c*. 20 cm x 15 cm) per brick uncovered, and the fact that most of the latter were half bricks.

Trench B

Trench B was located toward the western extremity of Henry's Mill, about 70 metres west of Trench A, and adjacent to a large, sprawling *Hydrangea*. Three main surface features were identified during survey: a 60 cm-high mound of brick and stone to the north-west, an alignment of small river cobbles two metres in length, and a further mound of clay and brick to the south-east (Figure 6; Plates 15-17). Surface artefacts included broken bottle glass, a short length of iron rail, and several rusted iron stove components. The 5 m x 5 m trench gridded to incorporate these features necessarily included a tree fern and two large, maturing specimens of manna gum (*E. viminalis*), which had contributed to the disturbance of the archaeological remains. Surface material also extended beyond the northern boundary of the trench.

The mound in the north-west covered an area of approximately 2.5 m x 2 m, with a scatter of stone and brick rubble extending a further 1-2 metres to the east and south. Revealed at the base of this mound was a large deposit of yellowish-brown clay and stone packing (**19**), at least two metres across, and up to 30 cm in depth (Figure 8). This deposit was built over a sterile layer of light brown clay and gravel. A number of large stones formed a border

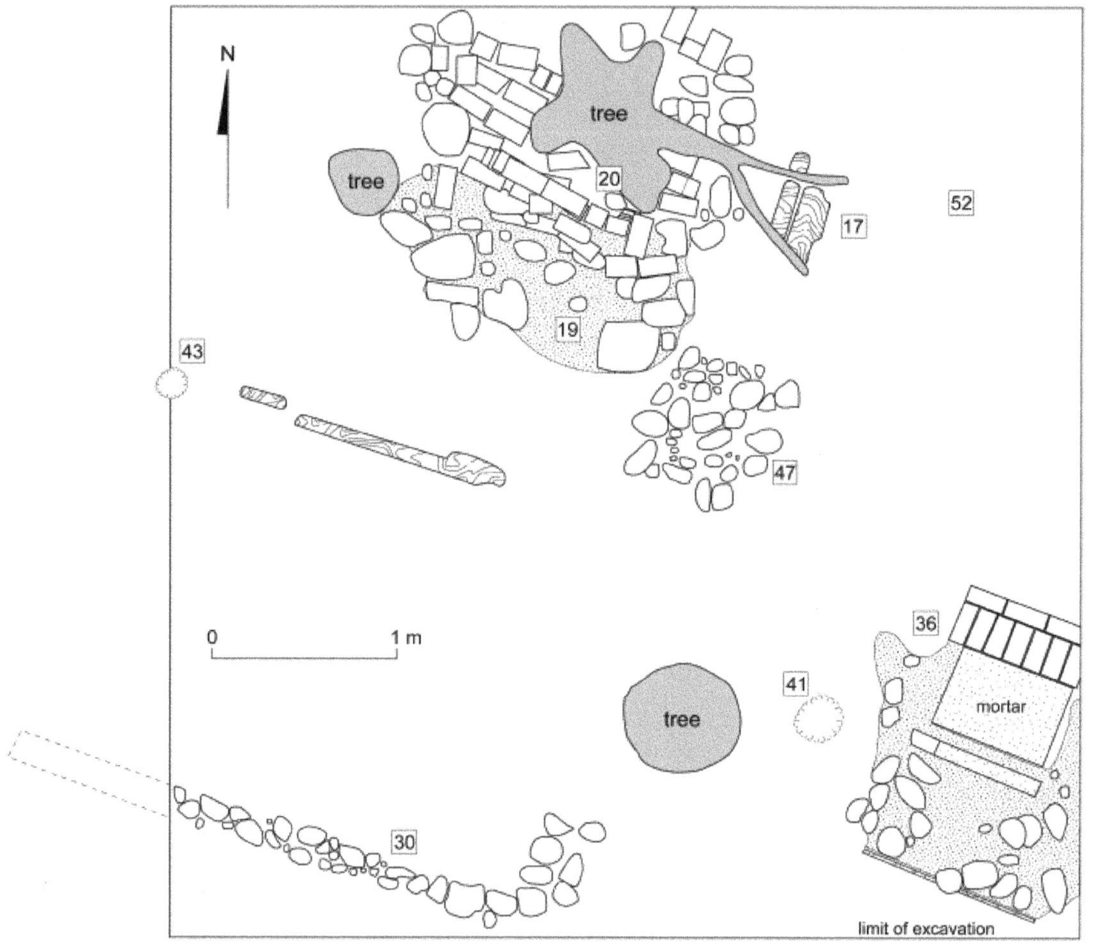

Figure 6. Plan of House B.

Plate 15. House B: Northern hearth feature (photo: author).

Plate 16. House B: Southern hearth feature (photo: author).

Plate 17. House B: Stone arrangement (photo: author).

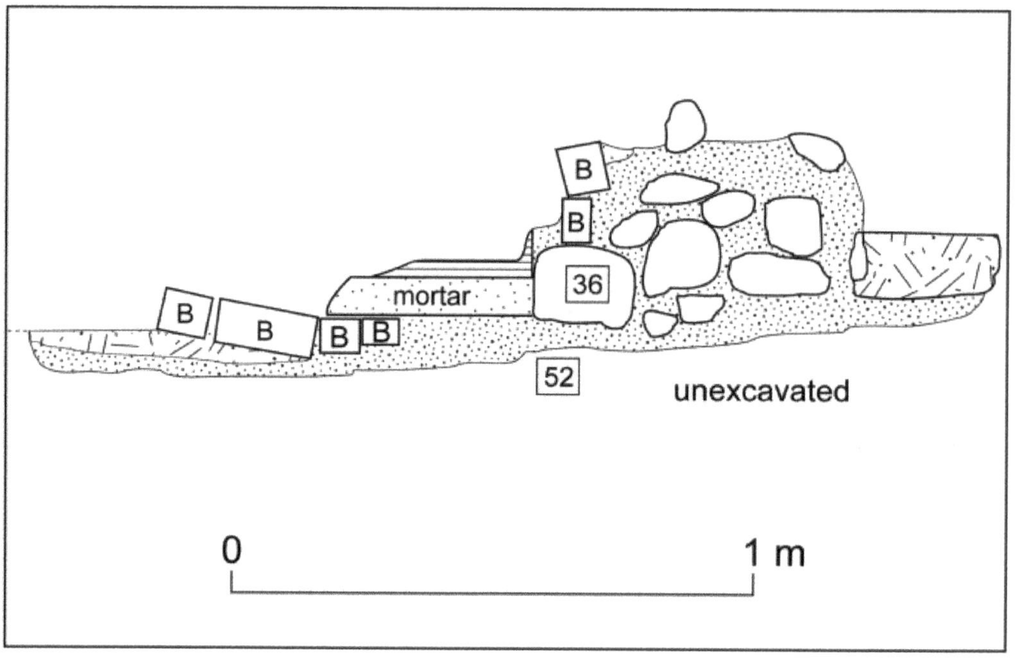

Figure 7. West facing section of southern hearth feature, House B

around this clay deposit, which appears to have formed a hearth and chimney foundation.

On top of this clay and stone deposit was a brick structure (context **20**), consisting of a 'wall' of mortared, bonded brick, two courses in width, at least seven courses in height, and one metre in length (Plate 15). Adjacent to the north of this wall was a brick platform, consisting of 3-4 rows of brick, two courses high. While each row of brick was bonded by sand mortar, the absence of mortar on the sides of the bricks indicates that either the individual rows were not bonded, or that the mortar has perished following the uplift caused by tree growth. A range of brick types were used in the construction of this feature, including examples from the Hoffman, Knights and Gartcraig brickworks. This structure appears to have functioned as an oven enclosure, similar in construction to context **36**, immediately to the south-east, along with contexts **9** in Trench A and **38** in Trench E.

The remains of probable wooden floorboards (**17**), preserved to a length of about 25 cm, were exposed to the east of the hearth feature **19** & **20**. A small fragment of linoleum was also found adhering to one of the boards. Possible floorboards or other structural timbers were located south of the hearth feature, the largest fragment being 1.20 metres in length. This area (**24**) was characterised by evidence of burning and fragments of corroded corrugated iron, which may originally have functioned as roofing or the internal lining of a chimney. In addition, an 80 cm x 80 cm arrangement of packed river cobbles (**47**) was uncovered adjacent to the south-east corner of the hearth feature **19** & **20**. Consisting of mostly angular stones up to 20 cm across, it appears unlikely that this 'cobbling' was intended to provide a surface for walking. Sectioning of this feature indicated that the stones had been placed in a natural shallow depression, and may have been used as deliberate packing, perhaps to deal with a drip zone.

The mound of stone and clay in the south-eastern corner of Trench B (**36**) had suffered much less disturbance by tree growth than the hearth feature **19** & **20**. This structure was built over a natural gravel layer (**52**), and consisted of two main features: a substantially intact brick platform or enclosure, and a clay and stone packing for a chimney foundation to the south (Figure 7; Plate 16). Numerous iron stove components were also scattered on the south-eastern surface of Trench B.

The soil deposit (29) located above and within the brick hearth enclosure contained a large proportion of artefacts typically associated with children, including a lead game piece, a tea set saucer and a piece of lined slate, along with several bricks, ash, rubble and metal fragments. The brick enclosure itself was constructed from several elements. The base, 62 cm x 41 cm, consisted of a layer of bricks, covered with a neat mortar finish up to 10 mm thick. There was no evidence of burning on the surface of this mortar. A brick backing wall, 72 cm wide and one course thick, rose three to four courses above the southern edge of the mortar base. Extending from this backing wall to either side of the brick and mortar base were two 'arms' of very sticky yellow clay.

Lining the interior edge of this clay and brick enclosure were corroded fragments of iron sheeting, originally providing additional protection and insulation. A small brick platform was constructed on the northern edge of **36**. It consisted of ten bricks, seven laid edge to edge, abutted by three more laid end to end, forming a 75 cm x 35 cm step in front of the hearth. These bricks were uniform in size and colour, machine-made, and all laid with the flat

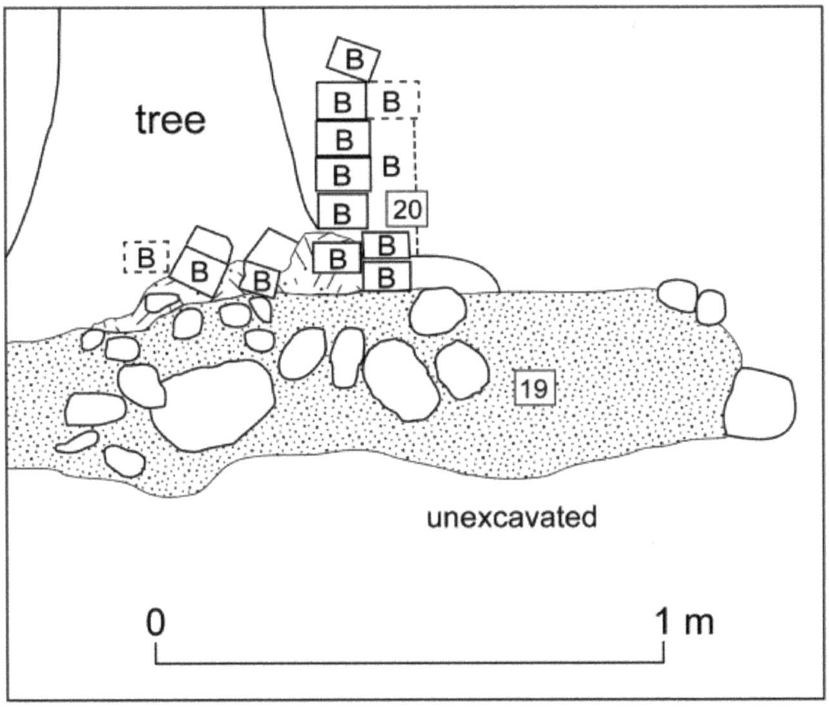

Figure 8. West facing section of northern hearth feature, House B

surfaces uppermost. Behind the brick step, base and back wall of **36** was a clay and stone packing which extended 110 cm x 60 cm. The southern side of this packing was edged by several lengths of timber, up to one metre in length. Brick structure **36** appears to have functioned as an oven enclosure, with the stone and clay packing providing support and insulation to the sides and rear , and probably acting as a chimney foundation.

Several other features were also exposed during the excavation of Trench B, including two possible post holes. The larger of these (**41**) was adjacent to the hearth structure **36**, circular in shape, and measured 20 cm in diameter and 20 cm in depth. The smaller post hole (**43**) was exposed west of fireplace **19** & **20**, of uncertain shape, measuring about 10 cm in diameter and 20 cm deep. Along with the remnant timbers identified to the east and south of the fireplace **19** & **20**, these post holes represent the only possible structural remains of the original house. The south-western corner of Trench B was dominated by an L-shaped alignment of stones (**30**), consisting of a single-coursed, double row of small river pebbles extending 3.20 metres WNW-ESE and 70 cm NNE-SSW, laid on the natural gravel surface (Plate 17). Its relationship to the brick and clay hearth features nearby is uncertain, but its most likely function was as a garden border behind the house. Alternatively, it may relate to a possible rear verandah or similar shelter over an external doorway associated with the cobbled area (**47**) between the two hearths, together with the two post holes (**41** and **43**).

In spite of the disturbance to the hearth feature **19** & **20**, and the minimal evidence relating to the architecture of the Trench B house, the relationship between the two main hearth features can still be assessed. While both appear to have functioned as oven enclosures, it appears unlikely that two would have been in operation, almost side by side, at the same time. It is known from documentary sources that individuals and families worked and resided at Henry's No.1 Mill for highly variable lengths of time, ranging from a few weeks to many years. The timber cottages in the north-western, residential part of the site were probably occupied by a number of households over the 24 years of the mill's operation. Given the typically rough construction methods employed, these houses presumably required maintenance and rebuilding to keep them habitable, a process which may have resulted on occasions in the rebuilding or replacement of hearth features with more suitable or functional arrangements.

In the case of Trench B, the hearth feature **19** & **20** may initially have been constructed as an oven enclosure, with a substantial clay base supporting a simple stone chimney foundation, opening to the north-east. Subsequent renovation involved the construction of a replacement enclosure (**36**) to the south-east. This feature was clearly purpose-built to enclose a kitchen oven, with the same orientation as **19** & **20**, utilising a uniform array of bricks which were neatly laid and rendered with cement mortar. The top, front, rear and base plates of an iron oven lying scattered on the surface indicate the width of the oven was 66 cm (*c*. 26 inches), and its depth was 38 cm (15 inches), dimensions which would have fitted snugly into the 68 cm x 44 cm space within the brick and mortar enclosure of **36**. The original hearth was remodelled as a simple open fireplace, utilising an array of bricks accumulated from a variety of sources. Several large iron artefacts were also noted on the surface (**6**) in the north-eastern part of Trench

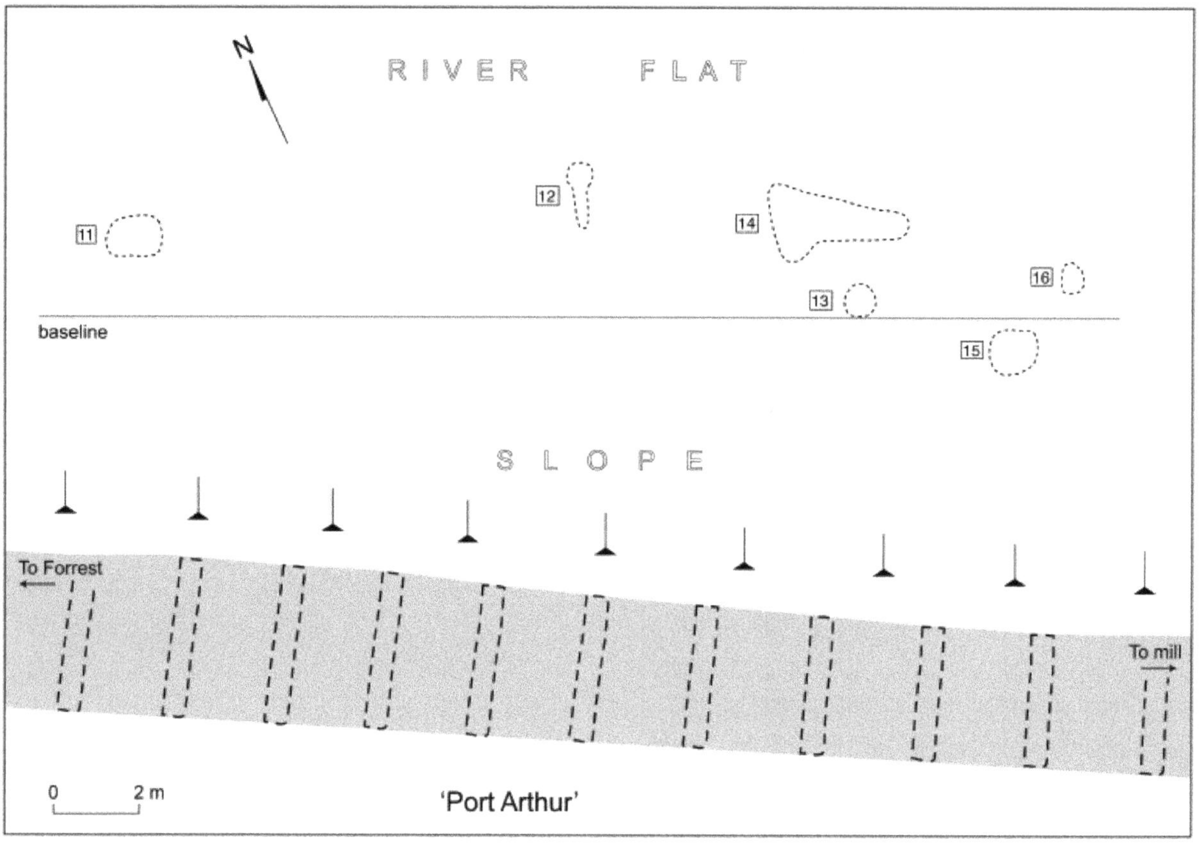

Figure 9. Plan of Dump C.

B, consistent with the suspension of pots over a fire. These included a length of heavy chain and iron ring, a 40 cm length of iron rail, perhaps used as a rough grate, and a metre-long pipe with flattened ends, inserted above the hearth.

Trench C

Trench C refers to a zone of discard immediately north of the 'Port Arthur' residential area of Henry's Mill. It lies five metres below the level of the tramline embankment, on the edge of the river flood zone (Figure 9). Initial survey of the area revealed scattered bottle glass in association with a series of small, shallow pits. The pits extended for 25 metres parallel to the line of the river, and ranged in size from 55 cm to 3.20 metres across, and generally 40 cm to 50 cm in depth. Initially these pits appeared to be the remains of toilets or the result of animal burrows, with the area representing a zone of intensive domestic discard. Close examination of the coarse sections visible in the pits, however, showed little indication of stratified material consistent with domestic middens. The irregular shape and size of pits, their visibility in an area of heavy weed growth, modern rubbish, and the surface clustering of artefacts around and within individual pits, suggested that bottle collectors had recently disturbed this area. Excavation in this area was thus avoided as unlikely to produce securely stratified material. Instead, a strategy of intensive surface collection aimed to provide a reference collection of artefacts, especially bottles, to aid in the identification of glass fragments from other parts of the site.

This process resulted in the recovery of 84 artefact fragments, including six ceramic sherds, 76 glass bottle fragments, a segment of barrel hoop and a fragment of slate. Bottles recovered were mostly large fragments, such as a neck and finish, shoulder and body or base and body. The area of Trench C appears to have been used as a general discard zone. Material lay in a context of secondary deposition, beyond its area of primary use (Schiffer 1987: 58). It lay adjacent to the tram track and the vicinity of the family cottages, in the flood zone of the river, an area largely unusable for most activities, and subject to periodic inundation and weed infestation. It is impossible to control for the origin of this material, other than most of it derives generally from the main residential part of the site.

Trench D

Trench D was a small 1 m x 1 m test pit in the area of the single men's huts (Figure 4). Excavation aimed to identify sub-surface features and recover a sample of material for comparison with that from the main residential part of the site. Intensive surface survey had indicated that this area was characterised by very dense scrubby undergrowth and minimal remains of cultural artefacts and features. With limited time and field resources, large-scale excavation was thus avoided in favour of a more restricted sampling strategy.

The test pit was excavated in the most promising location available, where a photograph of the site indicates a cluster of small timber huts beside a tramway branch line (Houghton 1975:57). Additionally, an iron tram wheel, piping and boiler binding, along with three whole bottles, had been discarded in the immediate vicinity. A shallow deposit, varying from 5 cm to 16 cm in depth, was excavated from Trench D (**25**), lying above a natural gravel and rock layer. Context **25** was a gritty, yellowish-brown deposit, thickly penetrated by root growth, which yielded 12 small fragments of bottle glass.

The area of the single men's huts generally appears to have been characterised by less extensive domestic debris than 'Port Arthur'. The construction of smaller wooden buildings left no discernible architectural traces. No evidence of remnant clay, brick and stone hearth features was identified. Upon abandonment of the site, the area was characterised by limited surface refuse and trampling, along with discarded machinery parts. Extrapolation from the quantity of artefacts recovered in Trench D to the remainder of the area of the single men's huts must remain speculative, although it is possible that material was discarded in a more concentrated fashion in other parts of the site. The flow of a shallow gully from the south through this area and the inevitably rapid deterioration of wooden structures may also have discouraged investment in substantial construction.

The small amount of material identified and recovered from this part of the site may also relate to the nature of single men's working lives. Their days were spent labouring in the sawmill or beyond in the bush, and most meals were typically taken in the boarding house. Some whiled away evenings in the billiard room, and many spent their weekends in Forrest or visiting families nearby. Many single men used their huts simply for sleeping. A simple, lightly built open fireplace was enough to heat the small structures and boil a billy, with no need for the more substantial hearth constructions found in 'Port Arthur'. Such huts were smaller and easier to dismantle upon abandonment of the settlement than the larger family cottages, resulting in fewer physical traces of occupation in this part of the site.

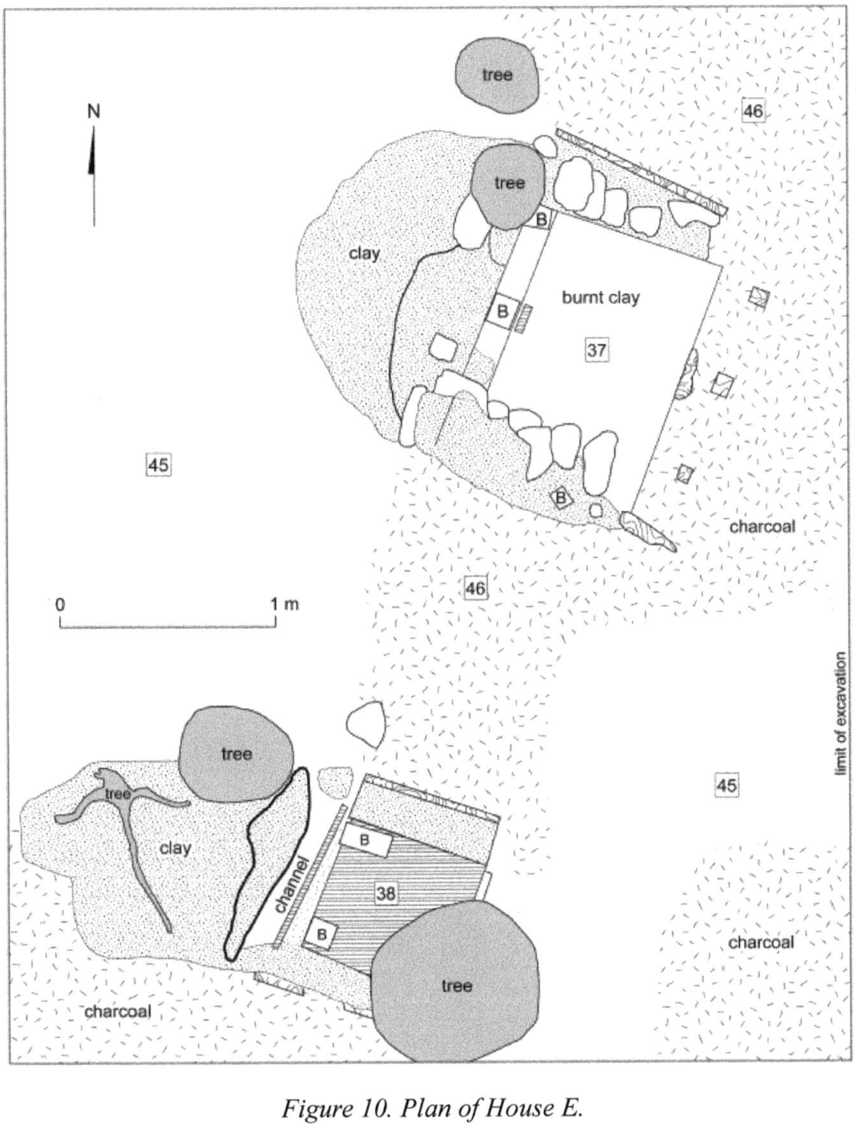

Figure 10. Plan of House E.

Trench E

Trench E was established 15 metres to the west of Trench A, extending four metres east-west and five metres north-south (Figure 10). The surface was dominated by two low mounds of gritty clay, from which were eroding bricks, stones and several lengths of iron rail. Four mature tree ferns and a prickly currant bush (*Coprosma quadrifida*) provided a substantial canopy over the surface features. Photographic evidence, together with the position of mature exotic pines, suggested that these features were likely to represent domestic debris, rather than the remains of the school building, which lay about 20 metres to the east.

The southern mound (**38**) in Trench E was irregularly shaped, with the clay, brick and stone having weathered and slumped after abandonment. It extended 120 cm between two tall tree ferns, and was preserved to a height of 60 cm. Exposed in the west of the mound was a large metal plate set vertically in the clay, measuring 77 cm wide, 45 cm high and 20 mm thick (Figure 11; Plate 18). Behind this plate lay a 15 cm-wide channel formed by a high protrusion of clay, apparently *in situ*. Immediately east of the mound lay several large pieces

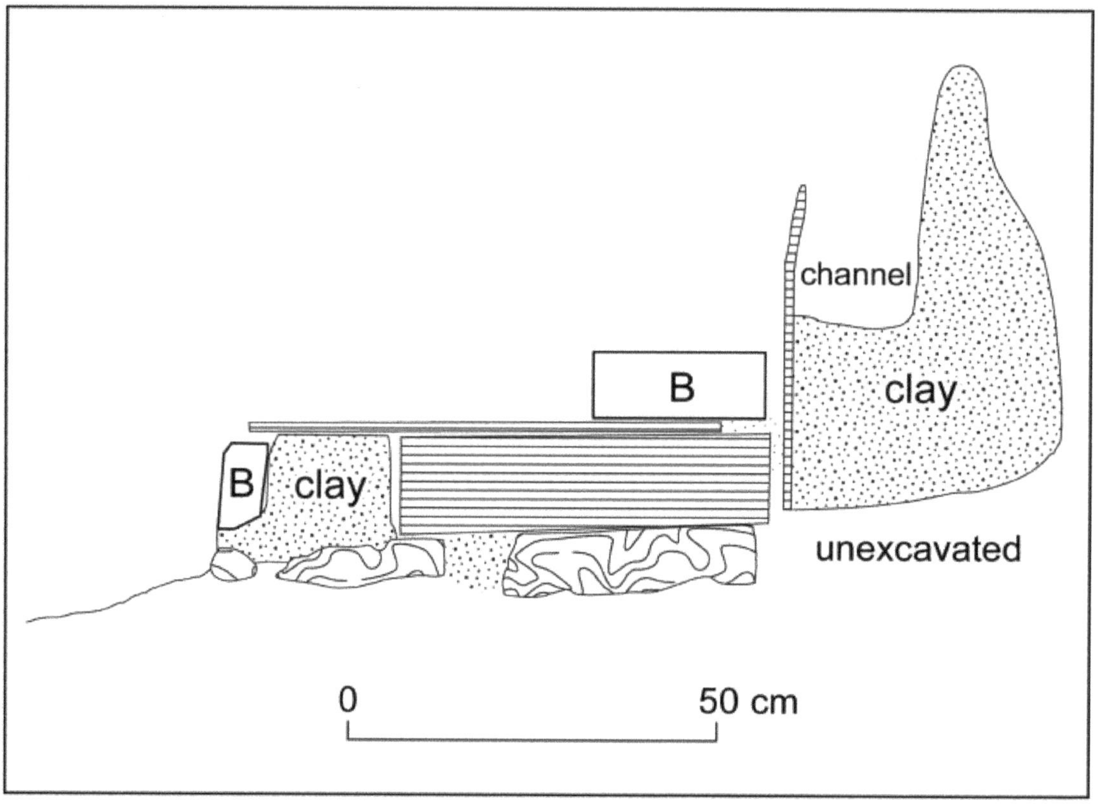

Figure 11. North-facing section of hearth 38, House E.

Plate 18. Southern hearth feature, House E (photo: author)

Plate 19. Northern hearth feature, House E (photo: author)

of discarded iron, including a partial oven top plate and a fire box base plate. Excavation involved clearing the mound of collapsed clay and brick to expose the structure below.

The foundation of **38** was a squared clay platform, measuring 110 cm north-south and 105 cm east-west. This platform was framed with timber on both the north and south sides, in a similar fashion to the hearth **8** in Trench A. Two iron plates, 50 cm long and 14 cm high, were inserted vertically into the clay, to support another iron plate, 65 cm square, which formed the base of the structure. On top of this base plate lay a shallow deposit of small bone fragments, above which a charcoal lens, containing glass bottle and metal can fragments, extended over the width of the plate. More clay was also packed beside each vertical plate for additional support. The clay later slumped to cover the charcoal and bone deposits.

One whole and one half brick were also exposed on top of the base plate, each flush with the large, thick iron plate set vertically on the western side of the iron base. The channel formed between this vertical plate and the high protrusion of clay behind it may originally have contained a low wall of bricks for support and insulation, before their removal and reuse. The presence of stove components on the surface of Trench E, and the solidity of the clay, brick and stone construction, indicate the likely function of **38** as an oven enclosure, with substantial clay packing at the rear and sides. The two bricks exposed at the rear of the base plate may have been used as stove footings, and/or have been associated with the post-abandonment re-use of the structure as a fireplace, resulting in the deposit of bone fragments and charcoal debris.

The northern mound in Trench E was irregularly shaped, up to three metres across and 80 cm in height. Removal of surface rubble revealed the structural remains of an open fireplace (context **37**; Plate 19). A rectangular bed of clay, 120 cm by 160 cm, was set in a wooden frame, built directly over the natural yellow-brown silty clay below. Burnt traces of the wooden frame survived on the north and east sides of the clay bed (Figure 12). The sides of the fireplace were constructed of flat stones, river cobbles and half-bricks, packed with clay mortar, while the remaining open side faced south-east. Clay in the centre of the hearth was friable and burnt red. An iron fire bar, salvaged from a boiler, 63 cm long and 13 cm wide, had been set upright in clay at the rear edge of the fireplace, and appeared to function as a structural support. A half-brick abutted the rear, western side of this fire bar, aligned with two more half-bricks in the north-west corner of the fireplace, one on top of the other. Around and behind these elements was more clay packing, forming a channel the approximate width of a brick. The rear edge of the burnt clay within the fireplace was aligned with the iron beam and the edge of the bricks. It thus appeared that the back of the fireplace was originally formed by a

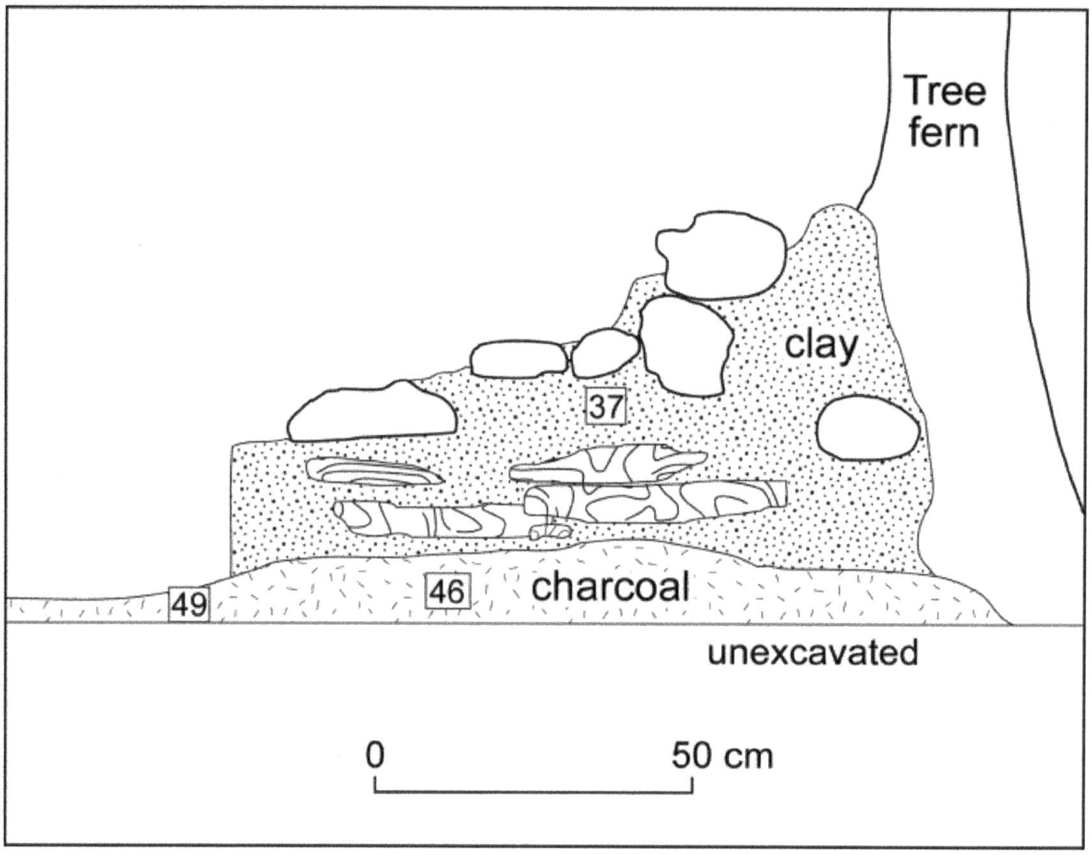

Figure 12. North-facing section of hearth 37, House E.

row or wall of bricks, most of which were subsequently removed, leaving only the clay channel to mark their position. Iron artefacts removed from the surface of the clay mound included two lengths of rail track (127 cm and 93 cm), and the notched, 60-cm-long shaft of a timber jack. The rail lengths may have been used as a crude grate to support fuel in the open fireplace.

A small test pit (**49**) was also excavated to examine a charcoal deposit (**46**) extending around the base of the fireplace (**37**). It exposed the probable remains of three burnt wooden floor stumps, each 45 cm apart. They were aligned straight and 10 cm from the front, eastern edge of the fireplace. The central, and best preserved stump, was roughly square, 9 cm by 7 cm in section, with a level, shaped upper surface, but sunk only a few centimetres into the underlying clay. A short remnant of burnt wood abutting this stump suggests floor joists were laid over the stump at this level.

Both hearth features **37** and **38** faced south-east and were aligned together, separated by a 1.40 metre gap. Abutting these two features and extending to all excavated areas of Trench E was a layer of charcoal (**46**), three centimetres thick, lying directly over the natural clay below. Several broken, heat-distorted glass bottles were also embedded in this deposit, along with numerous nails. It appears that a wooden floor, built on low stumps and joists, originally abutted both hearth features. Upon abandonment of the building, this floor was destroyed by fire. The charcoal deposit may also represent remains of a wooden chimney. Subsequent weathering of the clay packing around the hearths caused their partial slumping and collapse.

Site abandonment

The process of abandonment at Henry's Mill was a gradual one. Residents would have seen for themselves the mill's impending decline. The available log supply was dwindling, further away and of decreasing quality. Competition with sawmills in more favoured parts of Victoria, especially the Upper Yarra forests, constantly threatened market share. By the late 1920s, local newspapers were full of advertisements for motor cars, symbolising the changing technology of the era and the impending obsolescence of steam and animal power. The destruction of the mill shed in October 1927 thus precipitated a process of closure and abandonment which was soon to have become inevitable anyway.

The pattern of abandonment of the residential section of the No.1 Mill is informed by several conflicting sources of evidence. Conventional historical understanding of the site suggests that the mill settlement was dismantled, the huts and houses worth saving were broken into panels, and carted to the Noonday mill for re-erection (Houghton

1975: 69). Conflicting with this view is the contention of an informant, Mr Norm Gardner, who suggests that the school building was taken to Forrest and rebuilt into a house. The same informant recalls visiting Henry's Mill as a young boy, after its abandonment, and seeing old calendars left on the walls of standing houses which were unable to fit through the tram tunnel (N. Gardner 15 Nov. 1998).

Archaeologically, there is ample evidence that considerable remnants of the family cottages were left behind after the residents had departed. Apart from the substantial remains of hearths and iron oven components, there is also evidence for burning throughout Trenches A and E consistent with low intensity fire, possibly representing opportunistic campfire burning of remnant timber architecture. In addition, large quantities of nails recovered from each of the main excavation trenches appear to derive from wooden elements which either burned, or collapsed and rotted away in the typically damp site conditions. Cottages were constructed from timber sawn at the mill, especially manna gum (*E. viminalis*), messmate (*E. obliqua*), blue gum (*E. globulus*), mountain ash (*E. regnans*), and mountain grey gum (*E. cypellocarpa*). These timbers are not normally recognised for their durability in the ground (Hall, Johnston & Chippendale 1975).

There appears little reason to dispute the view that sawmill plant and equipment was salvaged for re-use, or taken to the Forrest railway station and dumped. Evidence for abandonment of the residential part of the site, and the subsequent formation of an archaeological deposit, however, is more complex. After the No.1 Mill shed was burnt out, logging operations continued in the West Barwon valley until March 1929, supplying logs to the Noonday Mill (Houghton forthcoming: 136). The post office was not officially closed until 31 October 1928, suggesting that a substantial body of workers continued to live at the site during final logging activities in 1928 and 1929, after the No.1 Mill itself had ceased operating. Quantities of artefacts on the surface may relate to this final phase of activity, when normal standards of cleanliness were relaxed in anticipation of abandoning the settlement in the immediate future (Schiffer 1987: 97). The roughly built cottages had stood at the site since 1904, and each had had several different occupants over this period. As residents did not own the cottages, maintenance may have been minimal, resulting in their gradual deterioration. While a few of these structures, including the school building, may have been worth saving, others appear to have been too dilapidated to salvage, and were abandoned to the elements.

The West Barwon watershed appears to have been variously effected by the impacts of logging and bushfire. Vegetation surveys of the valley in 1985 revealed an intricate mosaic of mature and overmature forest stands representing logging regrowth and regeneration following the bushfires of 1919, 1926 and 1939 (Dargavel 1995: 7). The presence of mature pines and oaks at the No.1 Mill, however, dating from the occupation of the site, appears to preclude the likelihood of intense bushfire at the mill. This contradicts the observation made by Blake (1973 vol.2: 956) that the '1939 fires obliterated all signs of mill and settlement', although it is acknowledged that the slopes east of the mill were burnt out in the fires of 1919, 1926 and 1939. Burning of the mill site thus appears to have been quite localised and of low intensity, with structures succumbing slowly to gradual deterioration.

Conclusion

The largest concentration of domestic debris at the No.1 Mill was located in the Port Arthur residential area. Excavation of Trenches A, B and E in this part of the site provided evidence of small structures, each featuring a pair of hearths to accommodate an iron kitchen oven and an open fireplace. Discussed in detail in Chapter 5, these structures are interpreted as wooden houses with timber and linoleum flooring, glass windows and split shingle or corrugated iron roofing. Trench C yielded no sign of architecture but abundant evidence of rubbish disposal. Trench D was a test pit in the area of the single men's huts which provided very little cultural material. No excavation was conducted in the industrial part of the mill site. Domestic refuse deposited around the mill dwellings and recovered during excavation forms the subject of the following chapter.

Chapter 4

Artefact analysis

Excavation at Henry's Mill aimed to recover a representative sample of domestic refuse from the main residential part of the site. The aim was to provide information on specific aspects of life at the settlement, including living conditions, consumption patterns and environmental responses. This material is described and analysed in the present chapter. The largest proportion of artefacts was recovered from House A, especially in the form of glass sherds, nails and metal fragments. Together with smaller quantities of similar materials from Houses B and E, this material makes up the great majority of artefactual evidence from Henry's Mill.

Each of the three excavated house sites represents the conflated discard of multiple households over about 25 years, formed via several processes of deposition. The bulk of artefacts derive from discard and trampling of domestic material around the houses, and items left behind in abandoned structures. Deposits for each house were essentially simple, amorphous sheet middens, obscuring the subtleties of separate household occupations. No rubbish pits were identified at the mill site, but the Trench C dump appears to represent a generalised discard zone beyond the immediate vicinity of the family cottages. Residents appear to have taken most of their personal effects with them upon departure, but left behind the house structures they had rented and fittings such as the iron ovens which they contained.

As material from each house site represents the conflation of successive occupations, some probably quite short, artefacts cannot be used to analyse individual 'households'. Further, although the names of many individuals from the No.1 Mill are known, the houses in which they lived are not recorded. These people thus cannot be connected to specific sets of material data. Analysis does not therefore attempt to compare and contrast different household assemblages. Differential sample sizes and a pattern of multiple, short-term occupations means that it is more appropriate to study the assemblage as broadly characterising a community group, one which represents an important segment of rural society in Australia during the early twentieth century.

Methodology

The classification system devised for processing material from the site had several purposes: to characterise the body of material in a comprehensive, meaningful and compact form; to facilitate storage and retrieval; and to identify significant artefacts for more detailed analysis. A broadly functional system of classification was employed, concentrating on material, form and specific attributes. Five activity sets were defined (architectural, domestic, industrial, personal, social), each of which was further subdivided into a range of functions. This approach aimed to permit analysis of material from the site at different scales, from broad generalizations about architecture, domestic activity, etc., to finer grained study of specific functions within these activity groups, such as food and drink consumption, education, medical support and personal hygiene. This classification system is based in part on the Parks Canada *Classification System for Historical Collections* (1992), although numerous other systems were also consulted (e.g. Adams & Adams 1991; Crook, Lawrence and Gibbs 2002; Davies & Buckley 1987; Hardesty 1988; Lawrence 1995; Praetzellis & Praetzellis 1990b; South 1977; Sprague 1980-81). The five activity sets for artefacts employed here are:

Architectural: artefacts originally created to define space for human activities. Functions: fastening; material.

Domestic: artefacts originally created to facilitate household activity and to meet the needs of people for comfort, sustenance and protection. Functions: food; furnishing; kitchenware; medicinal; storage; tableware.

Industrial: artefacts originally created to manage, harvest or collect resources, and to transform or modify raw and processed materials. Functions: fastening; hardware; hunting; tool.

Personal: artefacts originally created to serve an individual's personal needs as clothing, adornment, grooming, etc. Functions: accessory; clothing; hygiene; toy.

Social: artefacts originally created to facilitate social interaction between individuals and groups. Functions: currency; educational.

Artefacts recovered in the field were bulk bagged by fabric and brought back to the lab for cleaning, identification, cataloguing and analysis. Variables recorded on artefacts included quantity, weight, size, minimum number of individual (MNI) counts, completeness, colour, ware (ceramics), decoration technique, pattern and subsidiary functions. Makers' marks and place of manufacture were also recorded where possible. Nineteen separate artefact fabrics were identified during analysis: bone, brass, ceramic, copper, cork, glass, graphite, lead, leather, limestone, linoleum, masonite, metal (ferrous), plastic, rubber, shell, silver, slate and wood. More than 100 different artefact forms were identified among these materials, from 12 012 fragments, representing a minimum count of 3535 individual items, weighing in total 123.162 kg. Tables 1, 2 and 3 below present numerical summaries of these artefacts. Tables in Appendix C describe the artefact forms from each excavation trench by quantity, MNI and weight.

Form	Qty	MNI	Wgt (g)	Form	Qty	MNI	Wgt (g)
Architectural-fastening				split link	2	2	84
bracket	2	2	372	*subtotal*	23	22	5079
corner fastener	4	4	7	*Industrial-hardware*			
hinge	6	6	1139	boiler pipe	2	2	1039
nail	4481	2779	16 152	fish plate	2	2	1691
picture hook	1	1	1	rod	6	6	797
washer	12	12	75	*subtotal*	10	10	3527
wood screw	2	2	8	*Industrial-hunting*			
subtotal	4508	2806	17 754	rifle cartridge	18	18	11
Architectural-material				shotgun shell	15	15	52
brick	7	6	16 427	trap	1	1	757
flywire	28	1	1	*subtotal*	34	34	820
linoleum	4	1	2	*Industrial-tool*			
masonite	12	1	13	chisel	1	1	18
mortar	5	5	12	file	10	10	1857
window glass	81	4	163	punch/awl	3	3	371
subtotal	137	18	16 618	shoe last	1	1	1176
Domestic-food				wedge	1	1	355
bone	267	10	430	*subtotal*	16	16	3777
mollusca	4	4	6	*Personal-accessory*			
subtotal	271	14	436	bead	51	42	20
Domestic-furnishing				brooch	2	2	3
battery	3	2	104	clasp	3	3	5
clock weight	2	1	8	cuff link	1	1	2
figurine	11	2	26	earring	2	2	1
handle	7	7	140	spectacle lens	1	1	2
hook	3	3	380	tin	1	1	28
lamp glass	119	2	200	watch chain	1	1	4
lock	5	2	12	*subtotal*	62	53	65
mirror	1	1	1	*Personal-clothing*			
stove	20	2	3692	button	45	36	30
vase	5	1	121	eyelet	2	2	0.23
subtotal	*176*	*23*	*4684*	needle	1	1	0.29
Domestic-kitchenware				safety pin	1	1	1
bowl	15	6	71	scissors	2	1	20
handle	7	5	197	shoe leather	7	1	21
jug	5	3	53	*subtotal*	*58*	*42*	*72.52*
subtotal	*27*	*14*	*321*	*Personal-hygiene*			
Domestic-medicinal				teething ring	1	1	8
bottle	75	35	1346	toothbrush	1	1	6
glass slide	1	1	1	*subtotal*	*2*	*2*	*14*
jar	2	2	200	*Personal-toy*			
phial	14	9	127	doll	54	13	98
subtotal	*92*	*47*	*1674*	game piece	1	1	17

Table 1: Artefact forms by fragment quantity, minimum number counts, and weight.

Form	Qty	MNI	Wgt (g)	Form	Qty	MNI	Wgt (g)
Domestic-storage				marble	27	21	176
barrel hoop	1	1	329	model	4	4	43
bottle	3232	122	39 021	pendant	1	1	3
bottle stopper	11	11	101	toy tea set	9	6	38
can	132	7	535	*subtotal*	*96*	*46*	*375*
cork	1	1	1	*Social-currency*			
crown seal	13	12	52	coin	4	4	18
jar	36	7	427	*subtotal*	*4*	*4*	*18*
lid	5	5	51	*Social-educational*			
screw cap	4	2	7	pencil	53	31	85
tin key	2	2	9	slate board	12	4	90
subtotal	*3437*	*170*	*40 533*	*subtotal*	*65*	*35*	*175*
Domestic-tableware				*Unidentified-fastening*			
bowl	59	3	446	banding	35	13	569
cup	165	31	2027	clamp	1	1	24
dredger	8	2	21	strapping	12	3	238
egg cup	1	1	2	*subtotal*	*48*	*17*	*831*
lid	14	4	174	*Unidentified*			
plate	108	39	1660	bar	1	1	436
saucer	110	36	1634	disc	13	12	167
spoon	7	4	40	loop	5	4	49
teapot	5	3	65	plating	3	2	277
tumbler	7	2	51	*ring*	3	3	6
subtotal	*484*	*125*	*6120*	sheeting	52	7	423
Industrial-fastening				spring	2	2	4
bolt	2	2	885	tube	14	4	30
chain	5	5	2952	unidentified	2369		18 866
nut	2	2	56	wire	2	2	5
spike	12	11	1102	*subtotal*	*2464*	*37*	*20 263*
				Total	12 012	3535	123.162kg

Table 1 continued.

Fabric	Quantity	MNI	Weight (g)
bone	267	10	430
brass	11	11	51
ceramic	1487	190	24 934
copper	9	7	28
cork	1	1	0.85
glass	3794	275	42 512
graphite	1	1	0.11
lead	6	6	85
leather	7	1	21
limestone	7	7	35
linoleum	4	1	2
masonite	12	1	13
metal (ferrous)	6299	2956	53 678
plastic	9	8	27
rubber	2	2	9
shell	6	6	6
silver	1	1	4
slate	64	33	175
u/i	23	16	1146
wood	2	2	5
Totals	**12 012**	**3535**	**123.162 kg**

Table 2: Artefact fabrics by fragment quantity, minimum number counts, and weight.

	Quantity	MNI	Weight (g)
House A	7621	2263	59 121
House B	3149	918	36 360
Dump C	84	47	14 400
Sounding D	15	6	2152
House E	1123	292	10 596
Surface	20	9	533
Totals	12 012	3535	123.162 kg

Table 3: Artefact fragment quantities, minimum number counts and weight by excavation area.

Artefact Descriptions

Architectural – fastenings

Nails were overwhelmingly the dominant architectural fastening recovered from Henry's Mill, although various other items were also found in each of the three excavated house sites. House A yielded three corner fasteners, probably deriving from a wooden box or window frame, along with four metal hinges and eight iron spikes. Two metal brackets, three short lengths of heavy iron chain and a repair (split) link, two wood screws, two hinges and two iron spikes were recovered from House B. House E yielded one corner fastener, one length of chain and one iron spike.

Nails

The large quantities of nails recovered from each of the three main excavation trenches indicate the extensive role of timber in the construction of mill houses, and the ready availability of nails in such an environment (Tables 4 and 5). Nails from the No.1 Mill were generally in poor condition, often badly rusted and fragile, or heavily bound with accretion. As part of the general artefact processing, all nails were dried, brushed, sorted, counted, weighed and recorded. The nails recovered were almost universally steel wire nails, with either bullet (i.e. rhomboid) or rose heads (Varman 1987). In the United States, most nails sold by 1900 were wire nails (Wells 1998: 87). Both Adams and Varman note, however, that as British naileries switched to wire nails in the 1860s and 1870s, well ahead of the Americans, areas such as Australia, New Zealand and Canada had earlier access to wire nails (Adams 2002: 70; Varman 1987: 107).

Three broad subforms of nail were used to establish an index of size, and to identify the proportions of nails based on their likely function. An additional subform of 'unidentified' was used to deal with nail fragments: *short*: <51 mm (<2 inch); *medium*: 51-102 mm (2-4 inch); *long*: >102 mm (>4 inch). A total of 2779 nails were recovered from Henry's Mill, excluding nail fragments. This included 1207 short, 1358 medium, and 182 long nails. Twenty-four clouts, four tacks and three roofing nails were included in this total, along with one cut nail. Short nails were probably used for the final stages of carpentry, including panelling and joinery. Medium length nails were used in general construction for fastening weatherboards and floorboards. Long nails were needed to secure heavier structural elements, such as frames and joists. Tacks are consistent with the use of such floor coverings as linoleum.

Architectural – materials

Small quantities of house construction materials were recovered during excavation, including bricks, glass and linoleum. A total of 81 window glass fragments were identified from the three excavated house sites, along with small pieces of flywire from House B. The flywire may originally have formed part of a door screen or meat safe. House B also yielded small fragments of linoleum adhering to remnant floorboards, but no patterns were discernible.

Bricks

Due to their weight and difficulty of removal from the site, only a small sample of complete bricks (n=5) was recovered from the residential part of Henry's Mill, for comparison with those from other mill sites. Those retained were selected for their completeness and representativeness of the range of manufacturers observed on site.

	Short	Medium	Long	Other	u/i frags	Totals
House A	963	830	88	5	686	2572
House B	177	385	78	23	874	1537
Dump C						
Sounding D						
House E	67	143	16	4	135	365
Totals	**1207**	**1358**	**182**	**32**	**1695**	**4474**

Table 4: Nail subforms by quantity in each excavation area.

	Short	Medium	Long	Other	u/i frags	Totals
House A	2009	3500	2161	47	1929	9646
House B	303	1607	1282	63	2012	5267
Dump C						
Sounding D						
House E	110	520	287	10	312	1239
Totals	**2422**	**5627**	**3730**	**120**	**4253**	**16 152**

Table 5: Nail subforms by weight in grams in each excavation area.

Cat. No.	Maker	Location	Length (mm)	Width (mm)	Height (mm)	Weight (g)	Colour
HM1	AR& FB	South Yarra, Melb.	235	115	75	3100	mottled white
HM2	Knight's	Lal Lal, Vic.	236	113	80	3400	mottled white
HM3	Hoffman	Brunswick, Melb.	221	107	73	3200	white
HM4	Gartcraig	Ayrshire, Scot.	227	106	64	2800	white
HM5	Hoffman	Brunswick, Melb.	234	110	77	3900	red

Table 6: Bricks recovered from the No.1 Mill.

The two Hoffman bricks were produced at the Hoffman Brick and Pottery Works in Brunswick, Melbourne, some time after its establishment in 1870. Both were dry-pressed bricks, manufactured in steam-powered Bradley and Craven brick machines (Birmingham et al. 1983: 67; Stuart 1989: 29). The Gartcraig firebrick appears to have derived from West Ayrshire in Scotland (Gemmell 1986: 56), while the Knight's firebrick was produced between 1862 and 1898 near Lal Lal, 20 kilometres south-east of Ballarat (Jack & Cremin 1994: 62). The diversity of brick types, found both scattered on the surface of the site and incorporated into domestic hearth structures, indicates the expedient nature of their acquisition and use in the mill settlement. The reuse and recycling of bricks at Henry's is considered in greater detail in Chapter 8, 'Consumption'.

Domestic – food

A total of 271 faunal remains (bone and shell) was recovered from the No.1 Mill, weighing 436.3 grams. The relatively small size of the sample meant that it was analysed as a single data set, quantified in terms of identified specimens, representing a raw fragment count, minimum number of individuals, and by weight (Table 7). Specimens were identified to the lowest taxonomic level possible using reference texts and comparative skeletons of the main domestic and introduced taxa (Boessneck 1969; Getty 1975; Halstead & Collins 1994; Schmid 1972). Specimens which could not be identified were classified into one of three broad size categories (large, medium and small) signifying the size of the animal from which the specimen is likely to have originated. 'Large' represents animals of pig to horse size; 'medium' indicates dog to sheep size; and 'small' represents those animals of dog size or smaller. 'Unknown' represents those remains too fragmentary to assign to a size class. Specimens were further identified to skeletal element, including the component of the element represented. Evidence of butchery, burning, gnawing, breakage, the state of fusion and sex of specimens was also identified and recorded.

The fragmentary nature of the faunal material meant that 54.8% of the assemblage remained unidentified. Among the identified remains, the majority (60.7%) represented a homogeneous assemblage of domestic taxa, including cattle, pig and sheep/goat. The latter predominated in terms of both identified specimens and total fragment weight. Although cattle and pig remains were rare, the frequency of 'large' specimens among the unidentified remains implies that cattle were more abundant at the site than the identified remains alone would indicate. It is also possible that meat from larger-sized taxa, including cattle and pig, was brought to the site already defleshed. Nevertheless, on the available excavated evidence, the consumption of domestic taxa was mainly confined to sheep, supplemented to a minor extent by cattle and pig.

Among the wild taxa, the remains of rabbit predominated. Although the number of identified rabbit bones was second only to ovicaprid specimens, the rabbit's small size suggests it occupied a supplementary role in the diet relative to sheep. No evidence was revealed for the exploitation of native taxa in the assemblage, although specimens may exist among the unidentified bird remains. Overall, the faunal material indicates a clear bias toward the exploitation and consumption of domestic animals by the residents of Henry's Mill.

Analysis of ovicaprid skeletal elements also indicated poor representation of cranial and peripheral limb elements, implying that select cuts rather than entire carcasses were present in and around mill houses. The absence of juvenile

Species	Common Name	Identified Specimens	MNI	Weight (g)
Domestic taxa				
Bos taurus	Domestic cow	2	1	27
Ovis/Capra		69	3	186.2
inc. *Ovis aries*	Domestic sheep	4	2	42.8
Sus scrofa	Domestic pig	3	1	2.8
Subtotal		74	5	216
Wild taxa				
Oryctolagus cuniculus	European rabbit	38	3	15.3
Aves	birds	6	1	6.7
Mollusca	shellfish	4	4	6.3
Pisces	fish	1	1	0.8
Subtotal		49	9	29.1
Unidentified				
Large		22	-	82.6
Medium		122	-	107.7
Small		2	-	0.4
Unknown		2	-	0.5
Subtotal		148		191.2
Total		271	14	436.3

Table 7: Taxa represented in faunal assemblage from Henry's Mill, quantified by identified specimens, minimum number of individuals (MNI) and total fragment weight.

animals within the assemblage also suggests that sheep were raised and slaughtered off-site. Animal carcasses delivered to the mill were cut up in a small shed beside the boarding house for distribution to individual households (R. Henry 12 June 1999). The low representation of the main meat-bearing elements, such as the scapula, femur and pelvis, indicates an emphasis on the consumption of lower quality meat. Evidence of butchery was extremely rare, with only 3.3% of the assemblage revealing possible cut marks. This suggests that most meat bones were cooked whole in roasts, stews, etc, rather than as 'chops'. The lack of refrigeration and numerous flies at the mill meant that meat had to be eaten quickly before spoiling.

The majority of the faunal assemblage (70.7%) showed evidence of burning, with many of the burnt specimens displaying white discolouration. Bones typically become whiter as they are exposed to longer or more intense heat (Spennemann & Colley 1990: 57). Burnt bones from the No.1 Mill were thus exposed to high-intensity burning, although it is unclear whether this was in the form of high temperatures, prolonged exposure to heat, or a combination of the two. The high frequency of white bone nevertheless indicates that the remains were burnt following discard rather than during food preparation. Many may have been disposed of in fireplaces following a meal.

Meat consumption at the mill was thus based on sheep and, to a lesser extent, rabbit. Cattle, pigs and miscellaneous taxa such as molluscs and birds were utilised to only a small extent. Skeletal part representation suggests the acquisition mostly of medium to lower grade cuts of mutton. The very slight utilisation of wild resources indicates a pattern of food consumption firmly embedded in European traditions.

Domestic – furnishing

Fragments of fine, clear glass from House A represent the remains of two kerosene lamp chimneys, each with a 'corona' moulded rim (Cuffley 1984: 128; Woodhead et al. 1984: 61). Although Colac was connected with electricity in 1923, the grid was not extended to Forrest until 1955 (Houghton 1976: 14; McIntosh 1988: 98). In the absence of mains electricity at mill settlements, kerosene lamps and candles provided most domestic lighting.

Small generators, however, were available for household use by this period, employing wind-driven propellers or petrol motors, with power stored in sets of batteries (Brown, C. 1993: 173). The remains of two probable carbon core batteries were recovered from House A, consisting of a largely intact battery cell and a core from another cell. The dry chemical compound in a carbon core moved the electrical charge from the tin or zinc container to a core protruding from the centre of the cell. Both the core and the metal container were attached to a machine, producing a current running through the entire unit. Electrical power produced in this way was limited, as the chemical expired over time, so that the batteries were best used either periodically over a relatively long time (e.g. door bell) or steadily for a very short time (e.g. fire alarm; Dummer 1978: 54). Ric Henry recalls some residents at the mill using 12-volt accumulator batteries for electricity. His father, Ern Henry, had a car generator from an early model Dodge which he used to charge the batteries, for which he charged half a crown. Several hundred pounds a year were raised in this way, all of which was donated to the nursing home in Forrest (R. Henry 12 June 1999). The specific application of batteries at the No.1 Mill is uncertain, but it is possible they powered radios or portable lamps.

A small fragment of silvered mirror was also recovered from House A, along with the remains of two ceramic female figurines.

Domestic – storage

The late nineteenth century was characterised by a dramatic increase in the production and use of glass containers, as a range of technological developments transformed the traditional glass industry into a highly mechanised engineering activity (Baugher-Perlin 1982; Miller & Sullivan 1991). The use of semi-automatic glass-blowing machines in the later nineteenth century was followed by the introduction of the fully automatic bottle machine, patented in 1903 by Michael Owens of Toledo, Ohio. The machine developed by Owens gathered heated glass directly from the furnace; all processes of moulding and blowing no longer required direct human labour. By 1917 the Owens machine was producing half of the glass containers made in the United States, with the other half manufactured by hand blowers and operators of semi-automatic machines (Busch 1991: 119; Miller & Sullivan 1991: 104). Virtually all bottle manufacturing in Australia and Britain was automated by the 1920s (Boow 1991: 70). The standardisation achieved by bottle manufacturers by 1900 permits the broad identification of many bottles' original function on the basis of their form (Baugher-Perlin 1982: 259). The section below provides a brief outline of the main glass forms and closures recovered from Henry's Mill, followed by analysis of the quantity and nature of material from each excavation area.

Bottles

Beer/Wine Bottles

The simple distinction between amber-coloured beer bottles and green wine bottles is not readily sustained in the early twentieth century, as both were variously packaged in clear, light and dark amber, and light and dark green bottles (e.g. Arnold 1987: 91-127; Vader & Murray 1975: 55-6). They are thus grouped here together. The category includes bottles for light table wines and heavy fortified varieties (e.g. port, sherry), along with lager, ale and stout. Beer/wine bottles from Henry's Mill were generally made by Australian Glass Manufacturers or the Manufacturers' Bottle Company of Victoria. No labels were preserved, so specific brewers could only rarely be identified from embossed bottles. Australian wine bottles of this period were very rarely embossed (Arnold 1987: 95). Styles of bottle neck were typically either export (slightly bulging) or lager (sloping), with crown and ring seals predominating.

Condiment Bottles

Condiment bottles came in a large variety of shapes and sizes, and contained food additives or seasonings such as preserves, sauces, oils, extracts and essences. Food bottles of the nineteenth century were commonly aqua or light green in colour, whereas twentieth-century machine-made ones were usually clear (Baugher-Perlin 1982: 272).

Medicine/Pharmaceutical Bottles

Patent medicine bottles were manufactured in varying sizes and shapes for the storing, dispensing and sale of medications. They were generally of clear or aqua glass, oval or rectangular in cross-section, and up to 150 mm in height. Volumes ranged from *c.* 50-100 ml (1.5 to 3.5 fluid ounces). Many can be identified by inscriptions moulded into the glass. This category includes phials and pill bottles, which were smaller containers used to store and dispense liquid medicines and pills. They were frequently clear or amber in colour, and sealed by a cork or external screw cap.

Aerated water Bottles

This category includes all 'soft drink' bottles, such as aerated waters, ginger beers, hop beer, cordial and mineral water. Bottles used for these manufactures often had specific design features to deal with the internal pressure exerted by carbonation, including rounded bases and various types of patented closures. Aerated water bottles were generally circular in cross-section, and either clear, aqua, amber, green or cobalt-blue in colour.

Spirit Bottles

This category includes all identifiable spirit bottles, including square-profiled gin and schnapps, and clear whiskey bottles. Bitters are also included in this category, although it is unclear the extent to which this product was used as a medicinal or alcoholic beverage (Arnold 1987: 105; Vader & Murray 1975: 41).

Storage Jars

Food storage jars were developed in the nineteenth century, based on work carried out by Nicholas Appert in France and James Mason in the United States. Appert developed a process in 1810 of preserving food by packing it in an airtight container and boiling the container to destroy the bacteria. Mason patented his zinc screw cap in 1858, widely replaced after 1869 by a milk glass liner to prevent contact between the food and zinc cap. Preserving jars come in two basic shapes, square or round, with straight sides and a wide mouth, sometimes embossed with the manufacturer's name or symbol. These jars are most commonly in clear or aqua glass. This category excludes those beer and wine bottles recycled into jars.

Closures

Bottle manufacture in this period was also characterised by the development of a range of reliable closures, increasing the demand for commercial glass containers. Prior to 1850 the almost universal closure was the cork stopper (Lorrain 1968: 41-2). The Swing, or Lightening Seal was invented in 1875. It featured a porcelain (later vulcanised rubber) stopper anchored to the collar of the bottle by a heavy wire lever. The lever could be pressed down against the bottle neck to lock and seal the cap in place. This closure had the advantage of being reuseable, and was widely used for beer and carbonated drinks until after the Second World War. The Lightening stopper was also used for sealing wide-mouthed jars from 1890 until the present day.

James Mason developed an external screw cap for wide-mouthed bottling jars in 1858. His zinc screw cap, which forced a vulcanised rubber ring onto the glass lip of a fruit jar, was followed by a glass or porcelain liner to prevent seepage and contact between the zinc and contents. These wide-mouthed jars were widely adopted in Australia from the 1880s onwards (Boow 1991: 79-80).

'Club sauce' type stoppers were manufactured for small-mouthed commercial bottles in the late nineteenth and early twentieth century. These glass stoppers featured a circular top with a cylindrical tapered shaft. The top of the stopper had a large exposed area on which to emboss commercial marks (e.g. Holbrook & Co, Lea & Perrins). With few inherent weak spots, these stoppers are often excavated intact or missing only the shaft (Boow 1991: 70; Jones & Sullivan 1985: 152-3)

Codd's glass marble bottle was developed in the 1870s (Talbot 1974: 46-7). It consisted of an internal, pressurised seal between a smooth spherical ball and a ring of cork or rubber. The bottle was opened by pressing the ball down and breaking the seal, with projecting ridges inside the neck guiding the ball into a recess. Numerous variants of the Codd bottle were developed from the 1880s onward, but these were eventually replaced in Australia by the crown seal (Boow 1991: 78; Talbot 1974).

William Painter of Baltimore patented the crown seal in 1892. A cork-lined, crimped metal cap was forced down onto the sealing lip of the bottle, providing a reliable pressure seal for beer and carbonated drink containers. The widespread adoption of the crown seal in Australia only occurred in the 1920s, associated with the introduction of fully automatic bottle machines (Boow 1991: 69-70; Talbot 1974: 59-60).

A range of different bottle closures was represented in the material recovered from Henry's Mill. Crown seals were used for beer and aerated waters, while corks maintained their presence with wine (champagne), patent medicines and phials. External screw caps were used to seal jars and small pill bottles, while internal screws and Codd's stoppers sealed aerated water bottles. Club sauce stoppers generally derived from condiment bottles.

Analysis

A total of 3362 glass container sherds were recovered from excavations at Henry's No.1 Mill, weighing 38.701 kg and representing a minimum of 168 vessels. These figures exclude window glass, tableware, glass marbles and buttons, which are discussed elsewhere. Glass containers were analysed on the basis of profile and body elements, colour and manufacturers' marks. Cross-mends were made where possible. The main bottle forms identified were beer/wine, condiment, aerated water, spirit, patent medicine, ink, poison, perfume and storage jars. Identifiable specimens were manufactured in fully and semi-automatic bottle machines (Jones & Sullivan 1989:35-9). Apart from a number of largely complete bottles recovered from the surface of House A and Dump C, the fragmentary nature of most glass containers precluded the possibility of calculating the volume of contents. No evidence was recovered of paper labelling on bottles. Principal references used to aid in the identification and analysis of glass artefacts included Arnold (1987, 1990), Baugher-Perlin (1982), Boow (1991), Busch (1987), Fike (1987), Hutchison (1987), Jones, O. (1981), Jones and Sullivan (1985), Miller and Sullivan (1984), Shueard and Tuckwell (1993), Talbot (1974), Toulouse (1971), Vader and Murray (1979), Vermeulen (1979) and R. L. Wilson (1981).

House A

House A yielded the largest and most diverse array of glass forms recovered from Henry's Mill. A total of 2084 glass container sherds were recovered, weighing 12.852 kg, and representing a minimum of 79 vessels (Table 8). Eighteen condiment bottles were identified during analysis, both round and square in cross-section. They included a near-complete salad-oil bottle, four Rosella preserves, and two Holbrook bottles. Six club sauce type stoppers were recovered, one also embossed with Holbrook & Co. In addition, there were fragments of three clear, wide-mouthed storing jars, all with external screw necks.

A minimum of 19 beer/wine bottles were identified, including six near-complete bottles from the surface adjacent to the trench. Two clear glass sherds derive from a one-pint Irish Imperial Whiskey bottle, while a single amber sherd is consistent with a Cawsey Menck cordial bottle.

Twenty-five medicine/pharmaceutical bottles were also identified from House A, 11 of which could be partially identified by manufacturer. These included Mother Seigel's Syrup, Joseph Bosisto's eucalyptus oil (n=3), California Fig Syrup, Wood's Great Peppermint Cure, a Sarsaparilla bottle, Barry's Tricopherous for the hair and skin, and Greathead's Mixture. Among the five small pill bottles identified were three amber-coloured examples of Dr Morse's Indian Root Pills, while fragments of five medicinal phials were also identified. The shape of two other near-complete vessels suggests their likely function as ink bottles, while a small amber poison bottle featured moulded crosses and the words 'NOT T[O BE TAKEN]'.

House B

A minimum of 25 glass containers were recovered from House B, identified from 810 glass sherds weighing 5.766 kg (Table 8). These included three beer/wine bottles and a one-pint Irish Imperial Whiskey bottle. Two aerated water bottles were from J. J. Trait, and another from the Bollington Hop Beer Company, both of Geelong. Two condiment bottles were represented by a Jonathon Reeves pickle/sauce bottle and a club sauce stopper from Holbrook & Co. Eight medicine bottles included D. B. Jenner's Cough Balsam, Mrs Winslow's Syrup (for teething babies), and three medicinal phials. Three glass jars were also recovered, including a Mason's Patent, along with a single green perfume bottle.

Dump C

The glass material recovered from the surface collection of Dump C was notable for the frequency of large bottle fragments. A total of 76 glass container sherds was recovered, weighing 13.970 kg, with a minimum count of 41 vessels (Table 8). Thirteen of these were represented by at least half their original fabric. Twenty beer/wine bottles were identified, along with one whiskey and a probable bitters bottle. The seven condiment bottles included three Rosella preserves, a [Stonyfell?] olive oil bottle from Adelaide, and a Taylor Brothers Co-operative sauce bottle. The three patent medicines included a Chamberlain's Cough Remedy and two unmarked items. A Longmore's olive oil and Kruse's Prize Medal Magnesia were also identified on the surface during initial site survey. In addition, six beer/wine bottles showed evidence of having been cut down at the shoulder to form glass jars. These modified containers are discussed in Chapter 8 under the section 'Reuse and recycling'.

Sounding D

The small sounding in trench D yielded 12 glass bottle fragments, weighing 102 grams, and representing two beer/wine bottles, and an unidentified clear glass container. In addition, two complete beer/wine bottles from the Manufacturers' Bottle Company of Victoria (MBCV) were recovered from the surface adjacent to the trench, along with a near complete aerated water bottle from J. J. Trait of Geelong, featuring a variant Codd closure (Boow 1991: 77).

House E

House E yielded a minimum vessel count of 17 bottles, from a total of 377 sherds, weighing 6.011 kg (Table 8). These included seven amber and two green beer/wine bottles, a Rosella preserve bottle, and an aerated water bottle. Other glass containers included a small cylindrical

	House A	House B	Dump C	Sounding D	House E	Totals
Beer/Wine	19	3	20	4	9	55
Condiment	18	2	7		1	28
Medicine	25	8	3		2	38
Aerated water	1	3	2	1	1	8
Spirit	1	2	2		1	6
Ink	2					2
Poison	1					1
Perfume		1				1
Jar	3	3			1	7
u/i	9	3	7	1	2	22
Totals	79	25	41	6	17	168

Table 8: Glass container forms recovered from Henry's Mill.

medicinal phial, an Aromatic Schnapps bottle, a square sided storage jar, and a cobalt blue castor oil bottle. A half-length of barrel hoop was also recovered from House E, probably once part of a rainwater barrel.

Glass manufacturers' dates

Glass manufacturers' marks were generally located on or near the base of bottles. Identified marks mainly include examples from the Melbourne Glass Bottle Works, its successor from 1915, Australian Glass Manufacturers Ltd., and the Manufacturers' Bottle Company of Victoria (Table 9). In spite of dating discrepancies for these marks between Boow (1991: 176, 180) and Arnold (1985: 27), examples recovered from Henry's include specimens from the period 1900-1915, 1915-1923 and 1923-1930. An example from the United Glass Bottle Manufacturers of England derives from the Charlton works, near London, and post-dates 1921 (Boow 1991: 182; Toulouse 1971: 510).

Glass recycling

Bottles in this era were technically regarded as the property of the manufacturer, to be returned to the owner for refilling. Many from the No.1 Mill were embossed with such warnings as 'This bottle remains the property of...' or 'This bottle belongs to...'. It is clear, however, that many were never returned to the original supplier. This failure to

Form	Description	Qty	Date range	Reference
battery	dry cell	2	1868-*c.* 1920	Dummer 1978:54
bottle	Abbotsford Brewery	1	1904-1925	Deutsher 1999:148
bottle	AGM	8	1915-1923	Boow 1991:176
bottle	AGM	21	1923-1930	Boow 1991:176
bottle	Bollington Hop Beer Co	1	1903-1908	Deutsher 1999:118
bottle	MGBW	4	*c.* 1900-1915	Boow 1991:180
bottle	MBCV	22	1907+	Boow 1991:179
bottle	J J Trait	1	*c.* 1880+	Boow 1991:77
bottle	UGB	1	1921+	Toulouse 1971:510
brick	Hoffman	2	1870+	Stuart 1989:29
brick	Knight's	1	1862-1898	Cremin & Jack 1994:62
button	Volunteer Cadet Corps	1	1884-1906	Candy 1969:14
coin	half penny	1	1914	artefact
coin	sixpence	1	1908	artefact
coin	halfpenny	1	1913	artefact
coin	halfpenny	1	1921	artefact
crown seal		12	1892+	Boow 1991:69-70
doll	Armand Marseille	1	1899+	White, G. 1966:231
marble	ceramic	6	1884-1918	Randall 1971:103
marble	limestone	7	pre 1915	Randall 1971:102
plate	J & G Meakin	1	*c.* 1912+	Godden 1991:427
plate	Pountney & Co	1	*c.* 1889+	Godden 1991:507
plate	W H Grindley	1	*c.* 1914-1925	Godden 1991:294
plate	Ridgways	1	*c.* 1912+	Godden 1991:539
saucer	Triple Crown China	2	1891+	Godden 1991:11
shotgun shell	Eley No.12 London	2	1895-1919	Rutterford 1987:23
shotgun shell	Eley 12 Nobel	2	1924-1927	Rutterford 1987:23
shotgun shell	Kynoch	1	1895-1919	Rutterford 1987:23
stove	D & W Chandler	1	1893+	Sands & McDougall 1876-

Table 9: Dateable artefacts recovered from Henry's Mill.

retain bottles within the manufacturing and consumption cycle was a major problem for brewers and others selling products in glass containers in the late nineteenth and early twentieth century (Busch 1991: 116-17; Stuart 1993: 19). Bottle loss for the U.S. glass industry, for example, was estimated at around 65% in 1883 (Busch 1991: 117). George Wilmot, a manufacturer of aerated waters in Colac, offered a £2 reward in 1906 for information leading to the conviction of anyone found buying, selling or destroying bottles or syphons bearing his name (CH 5 Feb. 1906). Iain Stuart has noted a bottle at Henry's Nettle Mill which had had the statement of ownership deliberately removed (Stuart 1993: 19). Bottles were broken, discarded, reused by competitors or retained by housekeepers for their own use. In an isolated domestic context, where packaged products were less common than in cities, bottles may have had greater value for reuse in the home, and many from Henry's were reused and recycled for further storage (see Chapter 8, 'Reuse and recycling'; Busch 1991: 123; Hill 1982: 293).

Domestic Tableware

Five ceramic wares were identified among the material recovered from Henry's Mill, based on fabric colour, degree of vitrification, opaqueness, and type of glaze. A description of each is provided below. Ceramic terminology employed here is adapted from Brooks (2005), E. Cameron (1986), Majewski and O'Brien (1987), G. L. Miller (1991a, 1991b, 1992, 1993), Praetzellis and Praetzellis (1990a), Sussman (1985, 1997) and Webster (1971). The aim has been to achieve a consistent classification system, which avoids the confusion inherent in the great diversity of marketing terms used historically by ceramic manufacturers, and in modern archaeological analyses.

White earthenware includes a variety of ceramics whose fabric colour and texture ranges from a smooth white to a granular grey-white. White earthenware is generally dense, low in porosity, opaque and highly durable. A transparent alkaline or lead glaze provides a brilliant clear gloss. It was widely used for tableware in the later nineteenth and early twentieth centuries. The category includes wares known in the past under many names, including 'White Granite', 'Ironstone China', 'Semi Porcelain', 'Pearl China', 'Hotel ware', and others. White earthenwares were commonly decorated with transfer printing, hand painting, gilding, moulding, edge banding and decalcomania.

Porcelain is a dense, highly vitreous and transparent white-bodied ware. The fabric of porcelain vessels vitrifies completely during firing. Sherd break surfaces do not exhibit staining and are easily cleaned. Porcelain vessels were commonly decorated with hand painting, decalcomania, transfer printing, moulding and gilding. 'Chinese export' porcelains are thick-bodied wares with a granular grey-blue to white paste, usually decorated in a hand-painted underglaze blue design of a Chinese or pseudo-Chinese motif.

Stoneware is a dense, opaque, heavy pottery varying in fabric colour from buff to brown-black, depending on firing temperatures and clay impurities. Being non-porous, the body does not require glazing before use, although for cosmetic purposes a salt glaze on the exterior and a slip on the interior was often added. Stonewares are usually thicker and coarser bodied than earthenwares, and are most often used for storage and utilitarian containers.

Rockingham is a pale yellow or cream-coloured earthenware, covered with a streaked or mottled reddish-brown or yellow-brown glaze. Rockingham ware often has moulded decoration, and was commonly used for teapots, cups, jugs and ornamental vessels.

Earthenware includes a wide range of ceramics characterised by their variable colour and low firing temperature. Earthenware is opaque and porous, and requires glazing before it can be used for food preparation, serving or storage.

A total of 1415 ceramic sherds were recovered from Henry's Mill, with a total weight of 8.299 kg. All sherds were washed and sorted primarily on the basis of ware, shape, and decoration. Conjoins were made where possible. Minimum vessel counts were determined on the basis of form, rim sherd profiles, and decoration. Analysis of the ceramic assemblage revealed a minimum vessel count of 165 individual items. The bulk of these (n=112) derived from House A, with 33 items from House B, three from Dump C, and 13 from House E. Fragments of four vessels were also recovered from the surface. No ceramics were recovered from Sounding D. Overwhelmingly, ceramic vessels were of white earthenware, with 160 (97%) items represented by this ware. Only a single vessel each of stoneware (jar), Rockingham ware (teapot), and Chinese export porcelain (teapot), all from House A, were identified during lab analysis, along with two earthenware vessels (handle and rim). Fragments of children's toys recovered in the ceramic assemblage, including marbles, tea set pieces and doll fragments, have been excluded from this analysis and are discussed elsewhere.

Eleven simple vessel forms were identified: cup, saucer, teapot, plate, dredger, spoon, tureen, bowl, jug, jar and egg cup (Tables 10-16). The most numerous vessel forms identified were plates (n=39), saucers (n=36) and cups (n=31). Other forms were represented by far fewer vessels, and included three teapots, two dredgers, two spoons, six kitchen bowls, three jugs, two jars, one egg cup and one transfer-printed tureen lid. In addition, 39 vessels of unknown shape were identified from small fragments on the basis of distinct decorative motifs.

House A

The ceramic assemblage from Henry's was dominated by material recovered from House A. This area yielded a total of 1120 sherds, weighing 6.5508 kg, representing a minimum vessel count of 112 items (Tables 10 and 11).

This material constituted 79.15% of ceramics from the site by sherd count, 78.9% by weight, and 70% by vessel count.

The 21 saucers from House A were all of white earthenware, strongly utilitarian in appearance and manufacture, and were widely popular by this time. They lay within a narrow range of rim diameters (140-152 mm - 5½-6 inches), and ring base diameters (66-80 mm). A slightly depressed centre for holding a cup was also typical. While five saucers were undecorated, six featured painted floral or border band patterns, and another was moulded with a fluted design. The remaining nine saucers were minimally decorated with a thin (0.5-1.0 mm) gilt stripe on the interior body, three of which had an additional gilt stripe at the rim. No transfer-printed saucers were identified during analysis. Saucers were frequently represented by large sherds, four of which were cross-mended to more than one third of their original size.

Twenty ceramic cups were also identified from House A, and were, like the saucers, notable for their robust utilitarian manufacture, and glazed white earthenware fabric. Rim diameters ranged between 78 and 90 mm (*c.*3-3½ inches). Apart from one cup with a decal floral pattern, and three with painted rim bands, most cups from House A featured either a simple gilt stripe on the exterior body, 19-20 mm below the rim, or were undecorated. Cups and saucers from House A appear to represent vessels from several matching sets.

The probable remains of two teapots were also recovered from House A. One was a Rockingham teapot handle, and the other a Chinese export porcelain teapot lid with a single perforation, featuring an unidentified handpainted underglaze blue design. As each vessel is represented by only a single diagnostic sherd, shape identifications remain slightly conjectural.

The 29 ceramic plates recovered from House A were, like the cups and saucers, predominantly of robust, utilitarian white earthenware. They ranged in size from 160 to 260 mm in diameter (6-10 inches), and included dinner, breakfast and bread and butter plates. A variety of decorative techniques were employed, generally applied around the borders of vessels. Four plates were transfer-printed, including two of a set (8 inches and 9 inches) with an identical black foliate border pattern. Rim fragments of a large tureen lid with a transfer motif matching this pattern were also uncovered. There were no gilt plates from House A. Thirteen plates, however, featured rim and border relief moulding, mostly in a stylised leaf motif or wavy line pattern. The edges of these plates were typically moulded in a shallow ripple pattern. While some of these moulded white earthenware plates recall the 'Ceres' pattern outlined by Sussman (1985), the moulded plate designs from Henry's are less detailed and generally appear to depict foliate designs, rather than grain patterns.

A small quantity of glass tableware was also recovered from House A. This included two tumblers, one of which featured short mitres projecting from the base. A small glass plate was elaborately decorated with a press moulded rim, stippling and a sunburst motif on the base. Two lidded sugar bowls were also identified, along with a vase decorated with ten arched panels, a cupro-nickel teaspoon and two bone cutlery handles (Jones, O. 2000: 160-74).

Fragments of ceramic kitchenwares were generally small in size, and recovered in very limited quantities. Rim sherds of two cream-coloured, white earthenware jugs with relief moulding were identified, with diameters of 78 and 110 mm. Small fragments of a probable salt-glazed stoneware jar were also found in House A, along with a miniature white earthenware jar, and the rim of a brown-glazed earthenware bottle. Three ceramic kitchen bowls were also represented, each identified by small rim fragments, with diameters from 160-170 mm. Two metal pothooks were also recovered from House A.

House B

The ceramic assemblage recovered from House B was much smaller than from House A, but was also dominated by white earthenware cups, plates and saucers, generally with minimal decoration (Tables 12 and 13). House B yielded a total of 212 sherds, weighing 1.179 kg, representing a minimum vessel count of 33 items. In terms of tablewares, eight saucers were identified, two with painted pink border bands, two undecorated, and four with gilt stripes. Four cups were also represented, including one with moulded fluting and a pink border band, and another with a green transfer-printed image of a boy on a swing. Five plates, none matching as a set, were variously decorated with blue band painting or transfer floral motifs, while a 7-inch breakfast plate had a polychrome decal floral design. Two undecorated ceramic condiment spoons were also identified, along with an egg cup fragment. Kitchenwares were represented by a single cream-coloured jug, and two bowls. All three items were decorated with simple painted stripes on the exterior near the rim, below which were stamped medallion designs.

Dump C

Six undecorated fragments of white earthenware were recovered from Dump C. This material represented two cups and a kitchen bowl.

House E

Only small quantities of ceramics were recovered from House E, with a total of 66 sherds, 447.1 grams in weight, representing a minimum of 13 vessels (Table 14). All were of white earthenware, and with minimal decoration. Tablewares consisted of five cups, three of which were undecorated, while one had a thin pink stripe near the rim. Another was decorated with a green decal foliate motif, along with moulded and painted panels. Five saucers were also identified, again with three undecorated pieces, and another with a hand-painted overglaze flower. Two of the three plates from House E were transfer-printed with a blue floral and brown scroll design respectively. No kitchenwares or other vessel forms were identified from this area of excavation.

Maker's marks

The ceramic assemblage from Henry's Mill includes seven pieces with maker's marks printed on the base. Two saucers from House B feature a grey transfer-printed mark reading 'TRIPLE CROWN CHINA / [ENGLI]SH MAKE', with three crowns in the centre of the mark. A cup base from House E has a grey-green mark of 'W' inside the outline of a lidded jar, above the word 'AUSTRIA'. In addition, four flatware sherds from House A each include a green transfer-printed maker's mark. One reads 'J. & G. MEAKIN / HANLEY ENGLAND' and a crown motif. Another mark belongs to W.H. GRINDLEY & Co./ENGLAND, surrounded by a laurel motif, while a third includes the words 'BRISTOL SEMI P[ORCELAIN?] above the Royal Arms mark. The fourth reads 'RIDGWAYS FEDERAL / ENGLAND /

	earthenware	porcelain	Rockingham	stoneware	white earthenware	Totals
cup					20	20
saucer					21	21
teapot		1	1			2
plate					29	29
dredger					2	2
egg cup						
tureen					1	1
bowl					3	3
jug					2	2
spoon						
jar				1	1	2
u/i	1				29	30
Totals	1	1	1	1	108	112

Table 10: House A ceramics-minimum vessel counts by ware.

	undec	moulded	transfer	painted	gilt	decal	Totals
cup	5	2		2	10	1	20
saucer	5	1		6	9		21
teapot	1			1			2
plate	6	13	4	6			29
dredger				1	1		2
egg cup							
tureen			1				1
bowl			1	2			3
jug		2					2
spoon							
jar	2						2
u/i	3	1	9	15		2	30
Totals	22	19	15	33	20	3	112

Table 11: House A ceramics-minimum vessel counts by decoration.

	earthenware	porcelain	Rockingham	stoneware	white earthenware	Totals
cup					4	4
saucer					8	8
teapot					1	1
plate					5	5
dredger						
egg cup					1	1
tureen						
bowl					2	2
jug					1	1
spoon					2	2
jar						
u/i	1				8	9
Totals	1				32	33

Table 12: House B ceramics-minimum vessel counts by ware.

	undec	moulded	transfer	painted	gilt	decal	Totals
cup	2	1	1				4
saucer	2			2	4		8
teapot	1						1
plate	1		2	1		1	5
dredger							
egg cup	1						1
tureen							
bowl				2			2
jug				1			1
spoon	2						2
jar							
u/i	3		1	5			9
Totals	12	1	4	11	4	1	33

Table 13: House B ceramics-minimum vessel counts by decoration.

	undec	moulded	transfer	painted	gilt	decal	Totals
cup	3			1		1	5
saucer	3			1		1	5
teapot							
plate	1		2				3
dredger							
egg cup							
tureen							
spoon							
bowl							
jug							
spoon							
jar							
u/i							
Totals	7		2	2		2	13

Table 14: House E ceramics-minimum vessel counts by decoration.

	earthenware	porcelain	Rockingham	stoneware	white earthenware	Totals
cup					31	31
saucer					36	36
teapot		1	1		1	3
plate					39	39
dredger					2	2
egg cup					1	1
tureen					1	1
bowl					6	6
jug					3	3
spoon					2	2
jar				1	1	2
u/i	2				37	39
Totals	2	1	1	1	160	165

Table 15: Houses A, B and E, Dump C and surface ceramics-combined minimum vessel counts by ware.

ROYAL / SEMI-PORCELAIN / COMMONWEALTH / REG No.5716'.

The use of the word 'England' on three, and possibly five, of these marks generally supports their post-1891 manufacture (Godden 1991: 11). J. and G. Meakin were one of several Meakin firms operating at Hanley, England, in the late nineteenth and early twentieth century, including Henry (1873-6), Charles (1883-9) and Alfred (1875-), all of whom used the Royal Arms mark (Fisher 1970: 71-2). The design of the J. and G. Meakin mark dates it to post *c.*1912 (Godden 1991: 425-7). The Bristol Semi Porcelain

	undec	moulded	transfer	painted	gilt	decal	Totals
cup	12	3	1	3	10	2	31
saucer	12	1		9	13	1	36
teapot	2			1			3
plate	10	13	8	7		1	39
dredger				1	1		2
egg cup	1						1
tureen			1				1
bowl	1		1	4			6
jug		2		1			3
spoon	2						2
jar	2						2
u/i	6	1	10	20		2	39
Totals	48	20	21	46	24	6	165

Table 16: Houses A, B and E, Dump C and surface ceramics-combined minimum vessel counts by decoration.

plate was manufactured by Pountney & Co. of Bristol, sometime after *c.*1889 (Godden 1991: 507). The mark of W. H. Grindley and Co. was used from *c.*1914-25 (Godden 1991: 294). The Ridgways mark has no exact match in Godden, but is very close to an example post-dating 1912 (Godden 1964: 539). The Austrian mark, 'W', has proved more difficult to trace, and is not included in Cushion's (1980) overview of European pottery marks. Those ceramic vessels with dateable maker's mark are thus consistent with a manufacturing period spanning the 1890s to the 1920s (Table 9).

While the majority of ceramics found on nineteenth century Australian sites were of British manufacture, by the early twentieth century the origin of such vessels had become far more diverse. Examples from Henry's include a Chinese and Austrian source, while other possible locations include North America, Japan, France, Italy, Germany and elsewhere in Europe. Australian potteries may also have produced wares for local distribution and consumption. A base sherd with a Ridgways mark included the words 'Federal' and 'Commonwealth', and may have been manufactured specifically for the Australian market.

Hunting Material Culture

A total of 15 shells and 18 cartridges were recovered from Henry's Mill (Table 17). The shotgun shells featured a low-brass head, 21.5 mm in diameter, frequently with the internal paper wadding intact. Most were 12 gauge, manufactured by Eley of London or Kynoch of Birmingham. A single .410 shell head was also identified from House A. Sixteen .22 rim-fire cartridges were found in House B, and two in House A. Five shells preserve sufficient headstamp information to indicate manufacturing date ranges of 1895-1919 (n=3) and 1924-1927 (n=2) (Rutterford 1987: 23; see Table 9). Shells and cartridges were recovered mostly from within cultural deposits, with only a few found on the surface. The contexts of these items suggests significant hunting activity contemporary with the occupation of the site. This is confirmed by Ric Henry who recalls the use of rifles and shotguns at the No.1 Mill to shoot birds and rabbits (R. Henry 12 June 1999). Examples from the surface may derive from occasional visits by hunters to the site after its abandonment. Imported shotgun shells could be purchased in local stores at a cost of seven shillings per 100, roughly equivalent to a mill hand's daily wage (e.g. CH 12 March 1913). Smaller quantities of ammunition, however, were presumably also available. A complete rabbit trap with a four inch teethed jaw was also recovered from House A, consistent with the hunting of rabbits as a food supplement.

Tools

A range of largely intact tools was recovered relating to general carpentry and workshop activities. From House A these included a complete timber-splitting wedge, the tip of a chisel, and four files. One of the files featured the square edges and tapering width consistent with a mill saw file, used to sharpen saws and other cutting tools (Ross & Light 2000: 20). House B yielded six flat files and three punches/awls, as well as a 'foot' from a shoe last, apparently associated with the work of a cobbler. These may have belonged to a man called Monsborough, a distant relation of the Henrys. Monsborough had more or less

	12 gauge shell	.410 shell	.22 cartridges	Totals
House A	11	1	2	14
House B	2		16	18
House E	1			1
Totals	14	1	18	33

Table 17: Ammunition recovered from Houses A, B and E.

retired by the 1920s, but stayed on at the mill, repairing shoes and boots, and possibly leather machinery belts as well (R. Henry 12 June 1999).

Personal Accessories

Personal accessories included small items worn about the person for decoration or convenience such as jewellery, glasses and watch chains. Such items were found in limited quantities, perhaps because workers at the site rarely wore such accessories, and because those who did valued them and curated them carefully. These lost or discarded items may then represent evidence of occasional, discrete but visible expressions of display.

Two metal earrings were recovered from House A, while House E yielded a silver-plated cuff link and a brooch pin set with eight glass jewel pieces. A slightly larger array of such accessories was recovered from House B. These included a silver earring, a gilt brooch fragment, and a glass spectacle lens. This lens was the standard flat oval shape commonly advertised and worn during the nineteenth and twentieth centuries (Iacono 1999: 72). The lens appears to be a prescription piece, and thus worn not merely as a fashion accessory. A solid silver Albert bar was also recovered from House B. Alberts and half Alberts were long chains with a central bar which slipped through a waistcoat buttonhole, allowing a watch to be suspended and stored in the pocket.

Glass Beads

A total of 42 glass beads were recovered from Henry's Mill, with 12 from House A, 29 from House B, and one from House E (Table 18). The beads ranged in size from 1.7 mm to 14 mm in diameter, and were predominantly in shades of blue and green. They most likely functioned as decoration on garments, necklace parts or other jewellery items (Iacono 1999: 42).

House	Colour	Clarity	Size (mm)	Quantity
A	white	opaque	1.7	2
A	light blue	opaque	1.7	3
A	light blue	opaque	5.5	1
A	light blue	opaque	9	1
A	clear	translucent	8	1
A	clear	translucent	10	1
A	aqua green	opaque	8	3
B	light blue	translucent	7	3
B	light blue	opaque	5	4
B	light blue	opaque	7.5	1
B	light blue	opaque	8	3
B	light blue	opaque	9	1
B	light blue	opaque	19	1
B	medium blue	opaque	5	4
B	medium blue	translucent	8	1
B	clear	translucent	7	1
B	clear	translucent	8	3
B	clear	translucent	9	1
B	red	opaque	5	1
B	medium green	translucent	14	1
B	medium green	translucent	8	1
B	medium green	translucent	9	2
B	medium green	translucent	10	1
E	aqua blue	opaque	9	1
			Total:	42

Table 18: Inventory of glass beads from Houses A, B and E.

	glass	metal	shell	rubber	fabric	composite	Totals
House A	11	3				1	15
House B	7	7	2	1	1		18
House E	2	1					3
Totals	20	11	2	1	1	1	36

Table 19: Button fabric totals from Houses A, B and E.

House	Material	Colour	Fastening	Size (mm)	Quantity
A	glass	black	wedge shank	9.5	1
A	glass	pink	2 hole	10.5	1
A	glass	white	2 hole	10	4
A	glass	white	2 hole	11	1
A	glass	white	2 hole	12	1
A	glass	white	4 hole	10	2
A	glass	white	4 hole	14	1
A	metal		u/i	17	1
A	metal		2 hole	14	1
A	metal		4 hole	14	1
A	composite		2 hole	16	1
B	rubber	grey	4 hole	22.5	1
B	fabric		2 hole	9	1
B	shell	white	u/i	9.5	1
B	shell	white	4 hole	9	1
B	glass	white	2 hole	7	1
B	glass	white	2 hole	10.5	1
B	glass	white	2 hole	13	1
B	glass	white	2 hole	13.5	2
B	glass	white	4 hole	10	2
B	metal		alpha shank	17	1
B	metal		2 hole	12	1
B	metal		4 hole	16.5	2
B	metal		loop shank	8	1
B	metal		u/i	18	1
B	metal		u/i	13	1
E	metal		4 hole	17	1
E	glass	clear	2 hole	13.5	2
				Total:	36

Table 20: Inventory of buttons recovered from Houses A, B and E.

Clothing

Clothing remains from Henry's Mill were represented mostly by buttons, although leather fragments, shoe eyelets, a sewing needle and scissor arm were also recovered from House A, along with a safety pin from House B.

Buttons

A total of 36 buttons were recovered from Henry's Mill, manufactured mainly from glass and metal, along with shell, rubber, fabric and composite material (Tables 19 and 20). Fifteen derived from House A, 18 from House B and three from House E. White glass buttons ranged generally from 9-13.5 mm in diameter, and probably came from shirts and underwear. Having been mass-produced since the 1840s, such buttons were commonly available and were not gender specific (Iacono 1999: 53). Metal buttons, cut from a copper alloy sheet, were slightly larger, ranging from 12-18 mm in diameter, and originally fastened coats, vests and trousers (Lindbergh 1999: 51-2). Decoration on the buttons was limited to two examples of quality marks ('Best Ring Edge'), and one military insignia, with the inscription 'Volunteer Cadet Corps / Pro Deo et Patria / Victoria'. The buttons from the site are consistent with the use of shirts, underwear, trousers, coats, etc, and may well have been reused on several garments prior to their loss or discard.

Artefacts relating to personal hygiene

Only a small number of artefacts were identified relating to personal hygiene. House A yielded an amber-coloured plastic toothbrush handle made in England. Surprisingly, no items related to shaving were recovered, in spite of contemporary photographic evidence indicating that most mill workers of this period were clean-shaven (e.g. Houghton 1992: 88; McCarthy, M. J. 1987: 45). A white plastic ring recovered from the surface of House B is more difficult to interpret. The ring is 52 mm in diameter, with a visible joining section. A similar device from the Rocks, Sydney, 42 mm in diameter and made from bone, along with three others from the 'Little Lon' site in Melbourne, have been interpreted as medical or contraceptive pessaries (Karskens 1999: 173-9). Alternatively, they may have functioned as a baby's teething rings, as identical items made from plastic, rubber and bone were advertised, for example, in the Sears, Roebuck catalogue of 1897 (Sears, Roebuck 1897: 27).

Toys

Toy fragments recovered during excavation need not reflect the full range of materials used by resident children in their play activities. Children at Henry's swam in a dam on the river, played games with bows and arrows in the bush, ferreted for rabbits and fished. The toys recovered archaeologically were small and easily lost (e.g. marbles) or broken and beyond repair (e.g. doll's heads, tea set pieces), suggesting that the mill children, in spite of their isolation, were not necessarily deprived of manufactured toys. They also undoubtedly played with items of perishable material, such as rag dolls and skipping ropes (Davies 2005a).

Marbles

At least 21 marbles were recovered from Henry's Mill (Table 21). The majority of these were complete, ranging in size from 14 mm to 21.5 mm in diameter, made variously from glass, ceramic and limestone. Although machine-made glass marbles were becoming very common by the mid 1920s, all of the glass examples from Henry's Mill were hand-made. The marble collection is representative of late-nineteenth or early-twentieth-century children's items.

Clay earthenware 'commoney' marbles were popular between 1700-1900. Generally brownish in colour, they were mass-produced and inexpensive, and made up a large component of children's collections until *c.*1930 (Webb 1994: 14-15; Table 22). The two porcelain marbles from House A and House B originally had painted decoration which has been largely abraded away. Stoneware 'Bennington' marbles were popular between *c.*1870-1900 (Baumann 1999: 15; Carskadden and Gartley 1990: 57). Made from small clay pieces rolled into spheres, they were then coated with glaze and fired. Hastily made, they were often not truly round. A manganese glaze created the brown colour and a cobalt glaze resulted in the blue colour. Bennington 'eyes' are small circular marks that formed when they rested against others in the firing process (Randall 1971: 103).

Production of limestone marbles was based almost exclusively in Germany during the eighteenth and nineteenth century. The industry went into decline in the 1870s, but continued in production until about 1915. Two of the limestone marbles from Henry's Mill retain small facets on one or more sides, which are remnants of a cubic stage in the manufacturing process (Randall 1971: 102).

A number of glass marbles were also recovered. 'Clearies' were popular between *c.*1870-1930. There were also three German swirls, among the most common of hand-made glass marbles. The Lutz marble from House B was one of the most expensive glass marbles made. It included goldstone in its manufacture, and was made from *c.*1900-1915 (Baumann 1999: 44-45)

	House A	House B	House E
Clay 'commoney'		2	
limestone	3		
Bennington	2	4	
porcelain	1	1	
Glass 'clearie'		1	1
German Swirl	1	2	
Banded Opaque		2	
Lutz		1	
Totals	7	13	1

Table 21: Marbles recovered from Henry's Mill (all numbers are MNIs).

Dolls

Fragments of at least 13 dolls were recovered from Henry's Mill (Table 23). House A yielded six examples and seven came from House B, with those from the latter generally

	Price range	Range of production	Range of popularity	Total
clay 'commoney'	low	1500-1930	1700-1900	2
limestone	low	1700-1914	1740-1870	3
Bennington	Low-middle	1870-1914	1870-1900	6
porcelain	middle	1800-1914	1850-1900	2
Glass 'clearie'	Middle-high	1846-1930	1870-1930	2
German Swirl	high	1846-1930	1870-1930	3
Banded opaque	high	1846-1930	1870-1930	2
Lutz	high	1910-1915	1910-1915	1

Table 22: Marble types and values represented at Henry's Mill.

	Price range	Range of production	House A	House B
Penny doll	low	1850-1930		1
China doll	low	*c.* 1840-1914		2
Bisque doll	Middle-high	*c.* 1880-1930s	4	4
celluloid	low	1870-1930	2	

Table 23: Dolls recovered from Henry's Mill.

of higher quality. A small, complete 'penny' doll, 28 mm in height, featured a hairstyle that dates it to after 1900 (Rait 1989: 149-50). A glazed porcelain 'china' doll from House B featured crude spoon hands, suggesting a date in the mid-nineteenth century (Fainges 1990: 13), while another example from House A with poorly painted features and a 'low-brow' hairstyle dates the doll to *c.*1880-1900 (Pearson 1992: 43). Fragments of at least two celluloid dolls were also recovered from House A.

Fragments of bisque doll heads were recovered from both House A and House B. Bisque (unglazed tinted porcelain) was used by German and French manufacturers to make high quality dolls from *c.*1880 until the 1930s (Fainges 1991: 27). At least two examples from House A included 'Made in Germany' marks on the rear of the head, dating them to the early twentieth century (White, G. 1966: 202).

Other toys

A hollowcast figure of a horse and rider was recovered from House A, while a brass 'put-and-take' top came from House B. The latter were very popular from the nineteenth century to about the 1940s. Of the ten toy teaset items, more than half had some form of decoration (Table 24). There appears to have been a shift through time in the popularity of decorated tablewares for adults and toy teasets for children. The Australian colonial preference for colourful, transfer-printed table and teaware in the nineteenth century did not extend to the miniaturised forms, which were largely undecorated. By the turn of the century, however, advertised toy teawares were mostly decorated types, in contrast to the growing preference for undecorated or minimally decorated tablewares for adults (Ellis 2001: 59-60; Lawrence 2003a: 23-6). This pattern is also apparent at Henry's Mill, where the majority of adult table ceramics have little or no decoration, whereas the toy versions feature a variety of decorative motifs.

	ladle	lid	saucer	spoon	teacup
decal					1
painted		1	1		
relief		1			
relief (painted)			1		
Transfer-print					1
undecorated	1	1	1	1	
Totals	1	3	3	1	2

Table 24: Toy tableware items recovered from Henry's Mill.

Currency

Four coins were recovered from Henry's Mill. House A yielded two George V halfpennies, issued in 1913 and 1921. House B also yielded a George V halfpenny (1914), along with an Edward VII sixpence issued in 1908. The low value of the coins need not be indicative of the range of denominations originally in circulation, as pieces of lower value were more likely to have been mislaid and not retrieved (Johnson, A. W. 1999: 245).

Artefacts relating to Education

Writing slates and slate pencils are a key archaeological indicator of literacy and numeracy. They were used by children practising their schoolwork, by their parents preparing shopping lists, and by carpenters, masons and other tradesmen for marking a variety of surfaces (Davies 2005c). Writing slates have been in use since at least the fourteenth century (e.g. Chaucer 1894-97 II: 45), but the main era of their use began in the late eighteenth century, when developments in sea and land transport permitted the gradual expansion of slate quarrying in Wales. Common throughout the nineteenth century, there were calls by 1900 to remove slates from classrooms on the grounds of hygiene (e.g. Barcan 1988: 158; Lindsay 1974: 258). However, archaeological evidence from Henry's Mill indicates that slates persisted in use beyond this. The reasons lie in the economy and convenience of slates, the paucity of cheap paper as an alternative writing medium, and the persistence of nineteenth-century modes of teaching and learning into the first part of the twentieth century. Slates were still used in some Queensland schools in the 1960s, and elsewhere, such as India, writing slates and pencils continued to be produced and sold in the 1990s (Fulekar and Khan 1995).

Fragments of at least 30 slate pencils were recovered from Henry's Mill. These included 17 from House A, 11 from House B and two from House E, along with fragments of four writing slates. Slate pencil fragments varied from 17 mm to 62 mm in length, with individual pencils identified by a worn point at one end. Two examples from House A showed evidence of wear at both ends, suggesting reuse after one end was broken or became blunt through use. One example was brown in colour, where all the others were grey. A single graphite pencil fragment, 29 mm long, was also recovered from House A, along with two glass ink bottles.

Conclusion

A wide array of artefact forms and fabrics was recovered from excavations at Henry's No.1 Mill. The assemblage represents a broad sample of discarded domestic material from an early twentieth century rural settlement. Material from each of the three excavated house sites represents the conflation of multiple successive household occupations. The temporal range of dateable artefacts is consistent with the documented historical occupation of the site between 1904 and *c.*1928. Together with the excavated remains of household structures described in Chapter 3, this material forms the basis of analysis and discussion in the following chapters.

Chapter 5

Spaces and Structures

The physical spaces and cultural places of Henry's Mill provided opportunities for industry, movement, shelter and human interaction. Together they helped create, reinforce and alter social relations. The modification of landscape and construction of architectural features conveyed important social messages about social identity, roles and divisions within the mill community. This chapter considers these 'common places' of the mill, the structures and environments of ordinary working people (Upton and Vlach 1986: xiii). Analysis begins at the household level, as the locus of domestic activity for mill residents. Archaeological evidence indicates the importance of hearths as a structuring feature of family houses, with implications for the movement and activities of men, women and children. Within the settlement itself, industrial features such as the mill shed and stables are described, along with amenities for workers and residents, such as the boarding house, post office, store and school. The landscape of the mill environment is also described, with reference to the impact of logging in the area, and the dominant visual and aural qualities of the site. At the widest level, the relationship between the mill community and the outside world is considered in terms of transport and access along the timber tramline to Forrest and places beyond.

Housing

Houses at Henry's Mill were not built to last forever. Typical of accommodation provided at mill settlements throughout Victoria, they were constructed quickly and cheaply from wooden framing and unpainted weatherboards, freshly sawn at the mill. Flooring consisted of mill off-cuts, three inch by one inch boards, with gaps opening as the timber dried and shrank. Linoleum coverings, fragments of which were found adhering to a remnant floorboard in House B, were needed to prevent splinters from the roughly sawn boards. Charred timber in House E indicates the use of very low stumps, with no crawl space beneath the floor. Windows were small, either glassed-in or covered with a simple canvas blind. Splitters working in the forest nearby provided shingles for roofing, or else corrugated iron was brought in from outside. Brown paper or newspaper was pasted over hessian [burlap] cloth nailed to the inside of walls to keep out the draughts (Evans 1994: 109).

The general architectural pattern of family mill houses was based on a central front door under a verandah, opening into a main living area. Two small bedrooms made up the rear part of the house, often with a lean-to kitchen at the back. Rooms may have served several functions. Daryl Tonkin, a bush sawmiller, recalled his kitchen in the 1930s at Jackson's Track, north of Warragul, as 'the kitchen while we were cooking, the dining room while we were eating, the lounge while we were relaxing, the office while we were tallying the books, the bedroom while we were sleeping' (Landon and Tonkin 1999: 28).

At least eight houses were built in the small 'suburb' of Port Arthur at Henry's Mill. They clustered close together with the school building in their midst, facing north to the river, immediately south of the tramline back to Forrest. A teacher described them as 'absolutely destitute of architectural pretensions' (Branditt 1921: 13), while Ric Henry recalled them as 'pretty crude and pretty rough' (R. Henry 12 June 1999). Several other houses appear from school records to have been built on the opposite side of the river, but their exact location can no longer be identified.

Housing at forest sawmills was owned by the proprietors and built by the men they employed. At Henry's Mill, married workers with resident families paid rent of five shillings per week for their small houses, while single men lived in huts provided rent-free. At a time when landlords in the working suburbs of Melbourne charged 25 shillings a week for decaying three or four-room dwellings, married workers at Henry's Mill enjoyed very cheap housing (McCalman 1984: 110). A high degree of worker mobility, however, meant there was a regular turnover in the occupancy of each house. There was thus little incentive to maintain the houses beyond keeping them basically weatherproof, or perhaps adding a lean-to or woodshed for storage. Tenants were unlikely to spend much of their disposable income on the physical appearance of the structure, with the result that many houses were in poor repair by the later years of the mill's operation.

Archaeological evidence indicates that fireplaces were an important part of mill housing. A pair of hearths was identified in each of the three houses excavated. A clay and brick emplacement for an oven dominated House A, beside which was a small hearth of brick and iron rail forming a grate in an open fireplace. The former structure appears to have been associated with a kitchen oven, while the latter may have warmed a general living area. House B also featured a clay and brick oven emplacement, set two metres to the side and rear of a poorly preserved brick open fireplace. In House E, an open fireplace of clay, brick and stone was probably originally in the main living room, while another oven feature of timber, clay and iron plating was part of a rear kitchen. Oven enclosures in each of the excavated house sites were located immediately south of other hearth features, suggesting they were built as part of a kitchen at the rear of the north-facing houses. The general purpose open fireplace which the whole family could use, for cooking, heating, lighting and socialising, was thus supplemented by an iron box in another room intended only for cooking.

The one fire oven was widely adopted in Australia from the late nineteenth century. It generally consisted of a cast-iron box, set into a pre-existing hearth. An internal firebox heated both the oven compartment and pots placed on top. Such ovens had numerous advantages over the open fire for cooking. They were safer and easier to use, because the fire could be left enclosed and unattended for longer. Ovens were also more efficient, with more heat absorbed by the surrounding iron plates and insulation, and radiated back out for longer periods. Nevertheless, kitchen ovens often had problems, including cracked pipes, poor dampers and smoky flues. Stoves also had to be cleaned regularly, as they would rust if left dirty or undried, thereby creating another chore for women (Gollan 1978: 95). Although consuming a smaller volume of fuel, greater labour was required to cut and split the wood into the right size to fit the firebox, a task which typically fell to the sons of the families. Fortunately for residents of Henry's Mill, an abundant supply of waste timber from the mill was available for firewood.

One of the consequences of adopting stoves in the kitchen was a change in cookware. The heavy iron pots which had hung from hooks over the open fire were replaced with tinned or enamelled utensils that were lighter, easier to clean, and more convenient and attractive. Stoves also altered the way food was cooked and the kinds of food prepared. It was possible to boil potatoes, simmer a soup and bake a pie all at the same time, a combination next to impossible on an open hearth (Cowan 1983: 62; Gollan 1978: 96).

The size of kitchens in mill houses, however, remains uncertain. If large enough to accommodate a table for dining and space to prepare food, the kitchen may have persisted as the social centre of the home. Alternatively, if these kitchens were small, lean-to constructions, families would have been forced to gather in another room, away from the sights, sounds and smells of cooking, thereby isolating women and their work. The separation of ovens and fireplaces, however, may also represent a direct response to the climate of the West Barwon valley. Summer kitchens with ovens at the rear of houses helped to dispel excess heat. In winter, Dutch ovens, pot hooks and chains were used to prepare food over an internal open fireplace, while the rear kitchen was used to dry off wood, wet clothes and boots.

The limited space available meant that houses were generally built only a metre or two apart. Although they were individual, free-standing structures, thin wooden walls meant that conversations, quarrels, movements and belongings would have been readily discernible to neighbours and passers-by. The greater awareness of activities and voices may have added to the sense of community, as well as intensifying disagreements or disputes between neighbours. Even if noise and disruption became too much, however, the option to move away was very limited. Constructing a new house meant purchasing timber from the mill, clearing the vegetation and levelling a 'bench' on which to build, all in after-hours time. Given the difficulties involved, and the lack of utility for such a property when the mill eventually closed, meant that few families ever exploited the option.

A common feature of sawmill settlements in this period was the physical separation of family cottages from the huts of single men, a pattern which both created and reinforced social distinctions. Accommodation for the mill owner, married men and their families was often divided by a path, tramline or creek from that of unmarried employees. Single men's huts at the Toombullup Mill (1916-1929) near Mansfield, for example, clustered along the log tramline, while the married quarters were built on the opposite side of the mill, spread along a creek (Evans 1997: 70). At the Mississippi No.1 Mill near Warburton, a dozen family houses stretched along either side of a creek, while a row of huts for the single men were built beside the log yard (McCarthy 2001: 94). This segregation is evident at numerous other mill camps of the period in Victoria, New South Wales, Queensland and Tasmania, and gives some insight into the demography and social structure of mill communities. The physical separation of residential structures was a social phenomenon that was not necessarily enforced, but was understood as an appropriate arrangement of living space, predicated on marital, social and occupational bases.

In spite of the limited space within and around the mill site, personal interaction between resident women and unmarried men may not have been high. Working by day in the forest or at the sawmill, single men had little reason to venture several hundred metres along the tramline or riverbank to the 'suburb' of family homes. As well, women of the settlement, busy with domestic chores, may have had little need or opportunity to visit the industrial part of the site. There was neither church nor hotel at the mill, and most women appear to have used the mill store only rarely, ordering supplies instead from Forrest. Apart from posting a letter at the mill post office, or attending one of the occasional Saturday evening dances held in the schoolroom, most women apparently had only irregular contact with most unmarried workmen. The line of sight between family houses and the mill area was also partially obscured by the slopes and gullies of the site.

This physical separation between married employees with resident families and single men was blurred, however, by the accommodation quarters erected for married men immediately south of the mill shed, adjacent to the tramline (Houghton 1975: 57). A number of mill workers lodged their families in nearby townships such as Forrest, Barramunga or Barwon Downs, to whom they returned on weekends. It is uncertain whether these men paid rent for their lodgings, but they probably took meals in the mill boarding house. The married men's quarters represent a third form of residence at the site. The men were provided with accommodation quite separate from their unmarried co-workers, yet the communal nature of the quarters also indicated the absence of their wives and children.

Huts for single men were much smaller than the family houses. Each consisted of a single room about three metres by four metres across. They were constructed from sawn weatherboards or split palings, with a shingle roof and a large fireplace. The absence of any identifiable physical remains of these huts suggests that their external chimneys were built entirely of timber, with little or no stone lining of hearths. About 15 huts were constructed close together, west of the mill shed, largely within the course of a shallow, boggy gully draining from the south-west. To accommodate the number of single employees, two men were probably billeted to each hut, one a mill worker, the other a bush worker. Generally the bush worker would rise earlier in the morning to light the fire and feed the horses. The mill worker got up later, cooked breakfast and cleaned up after the bushman had left for work (McCarthy 1987: 58). Beds typically consisted of two bags with the corners cut out, with poles pushed through lengthwise and supported at each end on stakes.

The mill proprietor, W. R. Henry, also maintained his own timber cottage in the southern part of the site for when he or his family visited the mill (Houghton 1975: 57). In spite of intensive searching, no sign of the house was found during survey of the site. A former resident, however, recalled that the house was

> 'very rough, just sawn weatherboards, nailed together any old how, and the floor was just three by one [boards], just put together edge to edge, as they shrunk you'd hear them rattle rattle rattle' (R. Henry 12 June 1999).

When visiting, Ric Henry and his brothers slept under 'barcoo' blankets on the verandah with its canvas blind. Such 'sleepouts' were a regular feature of rural and suburban houses of the period. They provided ample, healthy fresh air and valuable extra sleeping space, as well as offering an architectural solution in the event of marital tension (Lake 1991: 48). The house had a pitched roof of galvanised iron, with two small bedrooms opening off a larger living room, the same pattern as the other workers' cottages. The walls were lined with hessian and newspaper and covered over with wallpaper. The connection of piped water from a gully behind the house probably reflects W. R. Henry's earlier training as a plumber (Houghton 1976: 28). Unlike the other family houses, the cottage never contained a kitchen or oven. Instead, meals were cooked over the open fireplace, an arrangement which may have suited the predominantly summer seasonal use of the dwelling. The authority of the Henry's as mill proprietors thus appears not to have been expressed in the construction of an elaborate on-site residence. Instead, its location at the highest point of the site, isolated from all the other huts and houses, provided enough spatial difference to separate employer from employees.

Amenities

Within months of Henry's Mill opening in 1904, the need to establish a primary school for the children became pressing. Parents were concerned their offspring were 'running wild' without a schoolroom and teacher to provide for their education. Children in Victoria had been entitled to 'free, secular and compulsory' education since 1872, so parents at the mill were keen to plead their case to the Education Department in Melbourne. The need for a school was argued with care. Their petition stressed the isolated location of the settlement, the local availability of pure water, and the willingness of the proprietors to erect a building. The likely initial attendance was cited as around 12 pupils, but likely to increase to 20, a margin well above the Department's typical minimum of nine children for establishing a school. It was also argued that the timber available to the mill was likely to last at least eight years, and that more married people with children would move to the mill settlement if a school was provided, thereby increasing pupil numbers even further. A District Inspector reported that the nearest existing school was at Barramunga, five miles away over 'very precipitous country' without roads, and that the parents were willing to accept even an unclassified teacher (VPRS 795/1927/3601, 9 Jan. 1905). This petition failed, and alternative arrangements had to be made, with parents sending older children to Melbourne and elsewhere for schooling. A subsequent petition in 1908, however, succeeded. The Education Department agreed to lease a building from the mill owners for a nominal yearly rental of £1, and appointed Leo O'Kelly as Head Teacher to open Otway Saw Mills No. 3601 in January 1909.

The school building was erected about 200 metres north-west of the mill shed among the family houses. A sketch plan prepared by W. R. Henry and two small photos reproduced in the school magazine *Forest, Lake, and Plain* show the school building faced the tramline that ran past 20 metres to the north (Branditt 1922: 34). The weatherboard structure was 26 feet long and 13 feet wide, with a gabled iron roof, internal pine lining and a dressed hardwood floor (Figure 13). For small windows provided only poor illumination. A wooden fireplace, pine door

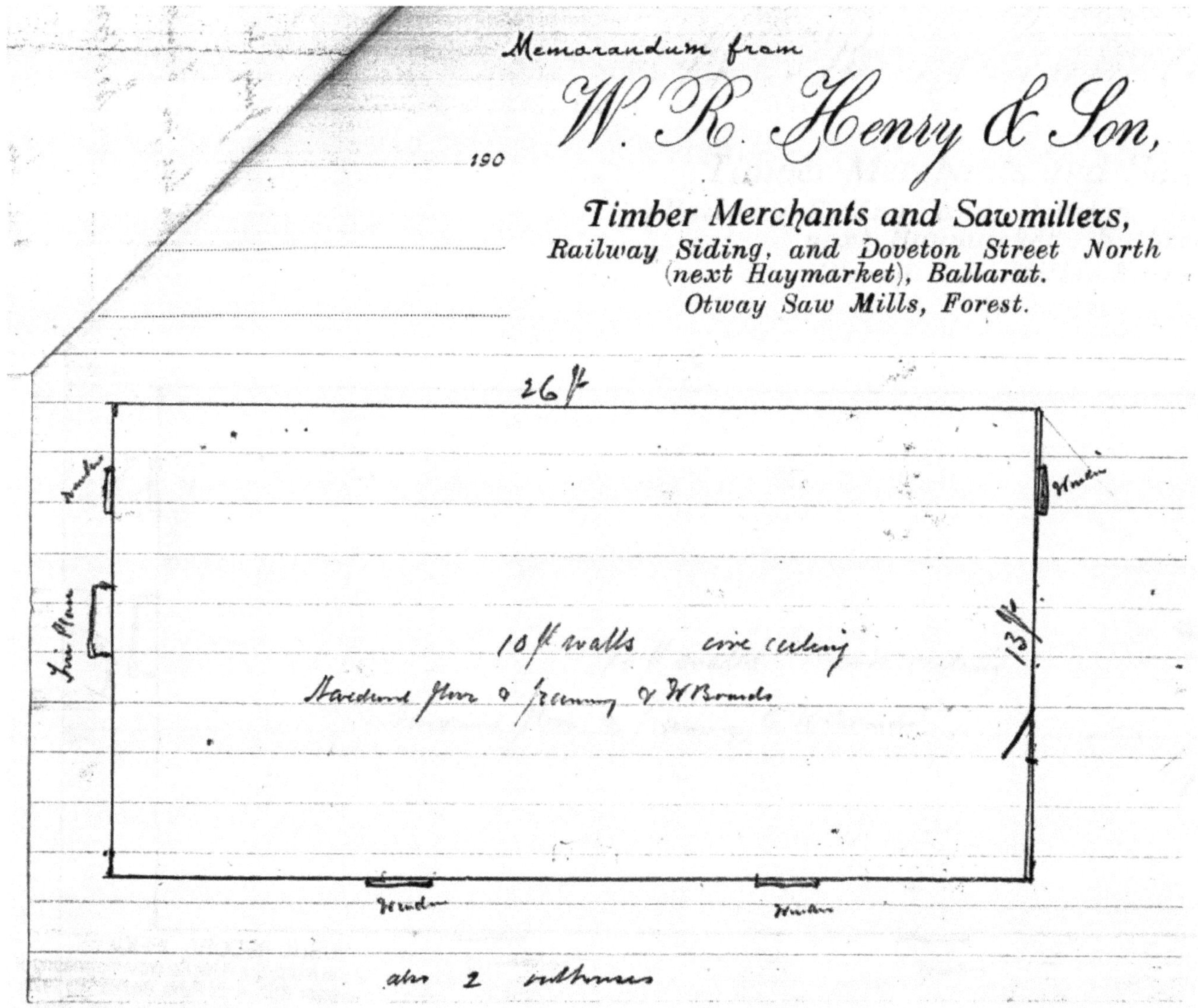

Figure 13. Construction plan for Otway Saw Mills School No. 3601

and a small ceiling vent of 40 square inches completed the structure. It was equipped with two outhouses and a small fenced playground, 22 yards square, adjacent to the tramline. A 600-gallon water tank was added in 1910, and in the following year a sand tray and observation case for Nature study were obtained. Inside were Department-issue desks and a portable blackboard, along with a Master's desk and stool, two book presses, an easel and a notation frame. Increasing enrolments by 1911 had some children sitting on boxes until more desks arrived.

Deterioration in the school building was apparent within a few years. This may have been due to poor design and construction, general neglect, weathering, daily wear and tear or a combination of these. By 1913, 100 bricks were needed to repair the back of the fireplace, and a few years later the playground fence was in disarray. In 1916, only eight years after its construction there were calls to replace the 'antiquated little school room' (Davies 2005b). The lighting, ventilation and floor space were by then regarded as inadequate and unhygienic by the School Committee, a group for many years comprising W. R. Henry as chairman, Cecil Turner as correspondent, the current teacher and one parent. The Education Department provided no maintenance allowance because it leased, rather than owned the building. It was not until 1923 that the Department agreed to increase the annual rental to £10 to permit the necessary improvements.

A small hut in the school yard, provided rent-free by the Henrys, served as a residence for Leo O'Kelly and his successor, Clifford Stanford. Subsequent teachers boarded with a family nearby, and converted the hut into a woodshed. Such basic accommodation was normal for teachers at bush sawmills. Blanche Murphy's hut at the Rubicon school, for example, was only nine feet square, built of roughly sawn weatherboards, and freely admitted 'rain water and mountain air' (VPRS 640/P1/1671/4 Feb. 1921). The teacher's hut at the Mississippi No.1 Mill was 'nothing much better than a Noah's Ark upon the waters. It is built nearly in the bed of the creek, and water flows right past it and nearly around it' (Blake 1973 vol.3:

453). The very basic teacher accommodation provided at sawmill settlements, however, was probably typical of that available to most teachers appointed to remote rural schools in Victoria during this era. Frank Tate, head of the Education Department in this period, referred to teachers' quarters as well named, mere 'vulgar fractions' of homes (Blake 1973 vol.1: 329).

Boarding houses were also an important feature of forest sawmill settlements. They were usually run by a mill worker's wife, by widows or deserted wives, providing the men with meals, but not accommodation. The boarding house at Henry's Mill was known locally as the 'Hash Foundry'. During the 1920s it was run by Martha Seebeck, and later by Eva Henrikson. Many of the men would cook their own breakfast and have their lunch and evening meal at the boarding house, with the charge docked from their wages. By 1924 the cost of board was 26 shillings per week, from a mill labourer's weekly salary of around £5.

Boarding houses fulfilled an essential role in the functioning of sawmill camps, and mill owners recognised the importance of providing single men with decent meals to improve the rough conditions of bush life. When G. W. Knott erected a mill at Wyelangta in 1911, he not only provided free timber to build the boarding house, but recommended to his men that they use the facility, and guaranteed their fortnightly board to the proprietress (Houghton 1992: 84).

Nellie Sands ran the boarding house at several mills in the Rubicon Forest of the Victorian Central Highlands between 1924 and 1936 (Evans 1994: 109). Her experiences were broadly typical of someone in her role in this period. Her day began around 5 a.m., cutting sandwiches for the men going out to work in the bush, before preparing breakfast for around 20 men. She took scones and cake to the mill for morning tea at 9.30, followed by lunch for the mill crew at noon, and afternoon tea at 3.30. Dinner then had to be provided for the men as well. She also had her husband and three children to care for, along with ordering her supplies, including sugar and flour by the sackful and butter by the box. She grew her own vegetables, baked all the bread required, kept a milking cow, and a brood of chickens for their eggs.

Former mill workers remember the fare provided at boarding houses in different ways. George Facey fondly recalls one mill in the Otways where the workers were fed 'like fighting cocks with fresh hot scones always available' (Facey, G. 1977: 20). At the Wye River mill, however, meals consisted of 'corned beef [for] breakfast, lunch and dinner, seven days a week' (Scales 1983). Some mills did not provide a boarding house, and the employees were responsible for their own food. Single men at Dumbrell's Mill, in far north-east Victoria, went home on Saturday afternoons and returned on Sundays with a week's provisions in a sack over their shoulder (Evans 1997: 97).

The boarding house at Henry's Mill was a simple rectangular wooden structure, with a sawdust heap on one side and horse stables on the other. Internally it featured little more than a single long table with benches for seating down either side, and a kitchen at one end. Maintaining sanitary conditions for the storage, preparation and serving of food was always difficult, given the swarms of summer flies and incessant smoke from burning wood waste at the mill. Tucked in next to the boarding house was a small shed for a butcher. The building was awkwardly propped up with poles and featured a sloping, skillion roof (Plate 11). Lacking refrigeration, meat butchered there had to be smoked or cured, or else consumed quickly before it spoiled. A wooden walkway extended from the log yard beside the mill shed between the butcher and stable, down to the river, providing access through part of the site which must normally have been littered with bark, mud and manure.

Henry's Mill was one of very few sawmills in the district to have a post office, complete with its own franking stamp, 'Otway Sawmills'. Operating out of the mill office, it opened as a receiving office in 1913, and was upgraded to a post office in 1921 (AA B5846/1 box 39). Letters were placed in a bag held on a wire frame, and delivered by the engine driver to the post office in Forrest six times a week. This facility ensured that workers and residents enjoyed almost daily interaction with the outside world, via letters, newspapers and deliveries.

A billiard room was also built by and for the employees of the mill. It was squeezed in between the stables, river and Forrest tramline. It contained two good quality Alcock tables, and provided the men with a place to meet and socialise. Little is known or recalled about this feature of the mill, other than the night when several men returned drunk from Forrest, lay down to sleep on one of the tables and ripped the cloth. A professional repair bill would have cost around £120, but Ern Henry patched it up and the tables continued in use (R. Henry 12 June 1999).

The Henrys also operated a small general store at the mill. Recalled as 'a real country variety store', it sold basic commodities such as tobacco, sweets, flour, sugar and tea. The profits on the store, however, must have been fairly marginal, as most of the single men took their meals at the boarding house and so needed to purchase little in the way of food. Married women prepared their families' meals from deliveries from Forrest, and so rarely needed to patronise the store either. The commercial choices of mill residents were thus extended by their access to outside supplies, without having to use the company store. The price mark-up on goods at the store is unknown, but it may have been similar to the practice at West Australian bush mills, where the cost was the Perth price plus ten per cent (TW 31 Aug. 1925).

The need for a public place, where all members of the mill community could gather on roughly equal terms, appears to have been given only irregular expression at Henry's Mill. The school building was used for religious services on Sundays as well as for occasional dances and school fetes. The boarding house provided meals to workers without resident families, but there is no indication it functioned as a meeting place beyond the arrival of hungry men three times a day. The billiard room may have been a magnet for men in the evenings, but it is unclear the extent to which women and children ever visited it. The nearby towns of Forrest and Barramunga, on the other hand, with their stores, pubs and other services, offered a more neutral, convivial alternative for people to gather on weekends away from the working environment of the sawmill.

Gardens were another significant feature of forest mill settlements. At Clark and Pearce's mill in the Rubicon Forest (1907-1939), for example, a Macedonian worker kept a market garden, while other residents maintained their own less extensive garden plots as well (Evans 1994: 109-110). Neatly cultivated household gardens were also noted at the Longwarry Mill in Gippsland in 1886 (McCarthy 1993: 38). There is also evidence for gardens at Henry's Mill. A stone arrangement excavated beside House B, for example, appears to have served as a border for a kitchen garden. Fenced gardens also lay behind several family houses. Exotic trees (Scots pines and oaks) and ornamental shrubs (*Hydrangea* and *Cestrum*) were planted and thrived, with little practical purpose beyond the immediate pleasure of their flowers, fragrance and appearance. The school also featured a garden in which the boys grew vegetables and the girls grew flowers. One of the mill workers, Oscar Henrikson, cultivated a large garden which helped supply the boarding house, while the smaller vegetable plots of other residents supplemented food supplies brought in from outside. The erection of fences around gardens created important spatial and social boundaries, and indicated the desire to claim, even notionally, some parcel of land in a context where neither individual nor group ownership of residential property was feasible. Gardens represented domestic comfort, and the establishment of a home. In an industry of high worker mobility, planting trees also demonstrated a commitment to remain at the mill for a season or two at least.

The milling landscape

Logging activities around Henry's Mill had a significant impact on the local landscape, with trees felled from the hillsides and creek gullies of the formerly untouched West Barwon watershed. Tramlines for log transport were erected up to seven kilometres to the east of the No.1 Mill. Although the District Forester described the method of logging in the area in 1925 as 'clearcutting', it is unlikely this matched the process of total coupe clearance typically practised in forest areas today (Ilie 1925: 11). Instead, logging practices were based on partial, or selective falling, with the sawmiller taking all the suitable trees available, and leaving those which were under or over-size, misshapen, over-aged or diseased. The labour-intensive nature of logging meant that no time was wasted on trees unlikely to provide suitable timber. Trees in inaccessibly deep gullies were generally not touched, or sometimes left for splitters to cut by hand into palings and shingles. Although Alexander Sanderson may have been glossing the impact of his own milling activities when he claimed there was 'still a dense forest left' after his crews had finished cutting an area, it is likely that the forest landscape was transformed rather than totally denuded (Minutes of Evidence 24 Feb. 1899). A patchwork remained of freshly cut slopes littered with bark and branches, scrub and bracken, mud, burnt areas, sapling regrowth, deep, untouched gorges and riverbanks lined with tree ferns.

The dominant colours of the No.1 Mill and its environs were grey, brown and green. For most of the year rain fell as almost constant drizzle, with only limited sunlight penetrating the clouds hanging low over the steep-sided valley. Joyce Harding remembers the weather nearby at Henry's Nettle Mill as 'wet...miserable...ghastly; it would rain for days' (J. Harding 21 May 1999). The steep hills flanking the No.1 Mill protected the site from most strong winds, and restricted the arc of sky visible from the bottom of the valley, reducing the hours of effective daylight. Smoke from summer bushfires added to the sense of isolation and threat.

Light was an important element in the lives of mill residents. The sun rose late and set early behind the steep hills of the West Barwon valley, on those days when rain, fog or mist did not obscure the sky completely. Small fixed windows, overhung by the sloping verandahs of family houses, admitted partial light, supplemented by that which filtered through cracks and gaps in roofs and walls. Most houses, however, were dim most of the time. At night rooms were lit by fires, candles and kerosene lamps, but care was needed when moving around outside in the dark to avoid mud, puddles, and unexpected logging debris and domestic waste.

A visitor to Henry's Mill was greeted by all the noise and activity of a sawmill in full operation. Steam engines puffed rhythmically, driving screaming saws through timber, and the roll of logs boomed in the loading yard. The blacksmith hammered at his forge, as mill hands chattered and laughed stacking timber. Dogs barked and children shouted at play, while their mothers clattered around kitchens to prepare meals for hungry families. In the forest, axes thudded into trees and wire winch ropes snapped along the ground. Teamsters urged bullocks, jingling in harness, to drag logs to the nearest landing. The squealing brakes of timber trucks echoed up and down the valley. The mill whistle, an ex-naval foghorn, signalled the times to rise, begin work, stop for lunch, and knock off at the end of the day. So loud was the whistle's blast that it was regarded as a danger to coastal shipping in Bass Strait, and Henry was eventually compelled to muffle it (Houghton 1975: 56).

Sawmillers in Victoria were normally granted a three acre license on which to erect a mill shed and fittings, along with a log yard, stables and stacking yards. The site for Henry's No.1 Mill was approximately 1.5 hectares in extent, chosen because it was the only wide and relatively level space in the valley south of the tram tunnel through the Barwon-Noonday spur. It was nevertheless a tight squeeze to construct the mill and associated buildings in the area available, and the slope on the site still represented an incline of 1:8. Most of the houses for mill families and the school were built at 'Port Arthur', several hundred metres to the north-west, on a narrow sloping ledge at the foot of a steep spur.

The site lay at the base of a distinct catchment system. Two small watercourses flowed through the main industrial part of the site, draining gullies from the south-east and south-west. The average annual rainfall of around 1500 mm was thus compounded by the added flow collected from surrounding slopes. With constant trampling over the years by around 100 people, along with horses, bullocks and dairy cows, the mill site must generally have been a muddy, boggy quagmire for much of the year. Wooden duckboards were used to make walking around the site easier.

Bullocks were grazed in the forest, along with a herd of about 28 dairy cows kept to provide milk for the mill residents. Mill horses, however, were stabled a short distance from the sawshed, just above the riverbank and between the billiard room and boarding house. Effluent from the 30 or more horses kept at the mill to haul log trucks would have contributed significantly to pollution of the river. Good quality horses were needed for the sawmilling industry because the rough and tumble character of bush work meant they had a shorter working life. Drivers of timber trucks were also responsible for feeding and grooming the horses. Although W. R. Henry commenced using steam locomotives for timber hauling from 1911, horse teams were retained for those steeper parts of the tram network higher up the valley. A dam behind the mill on the Barwon was deliberately formed where the sawdust pile sloped down to provide a swimming hole for the horse teams. It was used by the children as well.

Accessibility

Significant portions of the Otway forests, including the headwaters of the West Barwon River, remained unsurveyed at the beginning of the twentieth century (VPRS 1472/3, parish plans). The township of Forrest scarcely existed prior to 1892, when it was officially surveyed following completion of the railway branch line from Birregurra, which terminated arbitrarily beside the Barwon River in the northern foothills of the Otway Ranges. It was not until the late 1890s that the hamlet began to develop as a substantial settlement, and as late as 1897 some sawmillers of the area were forced to live in Barwon Downs, seven kilometres down the line, due to lack of available housing in Forrest (Houghton 1975: 7). The No.1 Mill was thus established in an isolated location in a remote district, ten kilometres south of the rail terminus at Forrest, with most employees forced to live permanently on site.

The tramline to Henry's Mill began in the Forrest railway station yard. An engine made a daily trip out to Henry's Mill, returning in the afternoon with timber, passengers and mail. The tramline provided a tenuous thread linking mill residents with local districts, regional towns and urban centres. It represented an umbilical cord with the outside world, a source of fresh news, faces and ideas. The line was owned, constructed and maintained by the mill proprietors, representing the firm's investment in and authority over the operating activities of the mill.

Due to awkward railway scheduling residents of the mill were at a severe disadvantage in terms of travelling to such regional centres as Colac or Geelong, even though they lived within a few miles of the rail terminus. A mixed train was scheduled to depart Birregurra at 10.25 am for the two-hour journey along the branch line to Forrest. On the return trip it left Forrest at 3.40 pm to arrive back at Birregurra at 5.25 pm (Houghton 1975: 11). Residents of Henry's Mill travelling to Colac thus had to make their way by foot or tram trolley to Forrest in time for the mid afternoon train, before connecting at Birregurra to arrive in Colac later in the evening. After conducting their business the following morning, it was impossible to return to Birregurra in time for the 10.25 am train back to Forrest and thence to the No.1 Mill. A simple shopping trip to Colac, 53 km away, thus took three days to complete.

Although the tramline provided the primary means of access to the outside world, a rough bush track also linked them through steep and isolated forest country to the township of Barramunga, eight kilometres to the north-west. The route, however, was regarded as virtually inaccessible. In its first year of operation, the school at Henry's Mill was worked on a part-time basis with the Barramunga state school. The teacher, Leo O'Kelly, was obliged to trek the distance each week on foot 'through excessively rugged and mountainous country', a journey 'attended by a large element of risk' (VPRS 795/1927/3601, 23 Oct. 1909). Even W. R. Henry acknowledged that in relation to nearby settlements his mill was 'in an almost inaccessible forest' (VPRS 795/1927/3601, 2 Oct. 1909). In such country, three miles an hour on horseback was classed as good travelling. Many of the 'roads' in the region, as late as the 1920s, were nothing more than tracks winding between the forest trees. Residents in the western Otways competed with those in the east in claiming their roads were the worst in Victoria. Ruts in some places were reported as almost four feet deep (CH 25 Sept. 1912). Although advertisements in local newspapers for motor vehicles became increasingly prominent during this period, emphasizing the revolution in personal transport under way, the notion of owning a car may have seemed futile to most mill residents when the nearest formed road was eight kilometres away. As well, a trip by coastal steamer to Melbourne, Geelong or other coastal centres was generally impossible, because it took more than a week for such a journey to be made. It was also expensive. A trip aboard the *Manuwatu*, which serviced the Otway coast between *c.*1900 and 1915, cost 15 shillings return from Melbourne (Pemberton 1979: 53). The isolated character of the region thus exacerbated the remoteness of Henry's Mill.

Conclusion

The cultural landscape created at Henry's Mill represented the response of workers and residents to the natural environment. In the early days of the mill's operation, the site was little more than a raw space carved from the dark forest, an open area in a sea of trees. Over time, as tramlines were extended and hillsides were logged and muddied, the landscape of the mill and valley underwent continual change. The settlement grew in local awareness, defined by the years it had existed, its increasingly dilapidated buildings, and people's varied memories and experiences of it.

Timber tramlines converged on the site from east, west and south. Their significance as the primary transport mechanism connecting the mill to the outside world was reinforced on a daily basis, with trucks continually shuttling from forest to mill and back. The single industry focus of the mill site required a substantial portion of the available space to be taken up by the sawmill, log yard, timber stacks and transport infrastructure. The rugged and broken topography of the area meant that relatively little flat ground was available for other amenities, resulting in the construction of houses and buildings in close proximity.

The rough timber construction of mill houses and buildings emphasised economy, expediency and ephemerality. Structures were never intended to last forever, built merely to provide basic shelter from the elements for machinery, operations, people, stores and animals. Buildings conveyed a sense of transience to owners and visitors alike. The architectural similarity of housing at the mill reinforced their construction by a single authority, the mill proprietors, over a short space of time, and obscured their occupation by distinct individuals and families. Nevertheless, the varied construction of fireplaces, and floor coverings such as linoleum, indicate that individuals reworked houses to some extent.

Social divisions at the mill received physical expression in the arrangement of living space. Huts for single men were clustered together between tramlines, the mill shed and a shallow gully. Houses for workers with resident families were located several hundred metres away at the foot of a spur. The schoolroom and yard nestled among them, reinforcing the family orientation of the area. Quarters for married men whose families lived off site were also quite separate, providing another level of social distinction at the mill based on marital status. The Henry family maintained their own small cottage as well. Although architecturally similar to other mill housing, its location at the highest point of the site, overlooking the operations of the mill, subtly emphasised the separation of employers from employees and the element of choice the Henrys enjoyed. In the following chapter the focus shifts from detailing the spatial context of the mill to characterising the people for whom Henry's Mill was both workplace and home.

Chapter 6

People

This chapter considers the people of Henry's Mill. It begins by examining the size of the mill population, and how this changed over the years of the site's operation. The number of children per household is also examined, and the implications this had for the role of women in families during this period. The ethnic background of workers in the timber industry is reviewed, along with the issue of internal migration between states and regions. These general patterns provide the context for exploring specific groups and individuals who lived and worked at Henry's Mill.

Mill residents tended to enter the historical record only when something went wrong and authorities became involved, in times of injury, death, illness or crime. From careful examination of these records it is possible to uncover personal, immediate responses to the physical and social environment of mill life. The second part of this discussion therefore traces the actions and lives of various individuals and groups at Henry's Mill, to explore the problems they confronted and the relationships they negotiated within the community. The correspondence of school teachers provides vivid insight into the range of responses articulated by a distinct group within the history of the settlement. This material, and the memories of Ric Henry, also helps to reveal the lives of the mill children, who formed a very substantial proportion of the community at Henry's. Attention is then paid to certain people and families in order to gain further understanding of the community. These include W. R. and E. A. Henry as the mill 'bosses', along with the extended McGlone clan, Cecil Turner, Oscar and Eva Henrikson, the working lives of Bernard and Gustave Seebeck, and Gladys Butcher and her family.

Population

The population of forest mill settlements varied according to a range of factors. These included sawmill capacity and the number of men needed to fall trees, saw logs and transport timber. Proximity to towns also dictated whether families lived on or off site, and where children went to school. The proportion of single men to married workers with families affected the size of mill populations as well. The smallest mills operated seasonally, and featured a handful of workers living at a camp during the week and returning home to town or farm on weekends. Hundreds of these 'spot' mills operated throughout Victoria between the 1850s and the 1930s.

At the largest settlements, however, several hundred people lived on site all year round, and such communities took on the air of small towns. The Anderson brothers, for example, established a sawmill at Barkstead, near Daylesford, in 1866, with the community of 150 people served by a post office, school, police station and store (Houghton 1980: 21). The Comet sawmill, near Wandong, directly employed 100 men during the 1890s, with two 'streets' of houses to accommodate all the families (Harvey 1992: 46). At the Ryson Creek mill in Gippsland, almost 200 people lived and worked at the site between 1891 and 1901 (McCarthy 1993: 46).

In between the largest and smallest mill settlements were many which employed 10 to 20 workers, and featured rough accommodation and a boarding house for meals. The Mackie Creek mill was typical of these settlements. Located six kilometres north of Henry's Mill, it started operation in 1908. Recalled as 'a regular little community in the bush', it had eight four-room houses, an office, 12 huts, a smithy, stables and cookhouse. Many men had gardens with potatoes and other vegetables along the banks of the creek. The company also owned five houses in Barwon Downs for drivers, the blacksmith and other mill men (Moulds 1991: 35). There was a similar arrangement at King and McCulloch's mill on Wilson's Promontory, where familes and single men comprised a group of forty people between 1903 and 1906. The settlement featured huts and houses, a community hall, office, workshops and boarding house (Lennon 1988: 432).

Around 100 people lived at Henry's No.1 Mill by 1909 (Houghton 1975: 56). This fluctuated according to the number of families present, the health of the timber industry and the willingness of workers to reside at the mill. Families comprised approximately half of the mill population. Households of the 15 or so married workers with resident families averaged 2.25 children each; the mode was three children (VPRS 795/1927/3601, 9 Jan. 1905). Larger families, however, were still reasonably common, and included those of Claudia and Thomas McGlone (eight children), Catherine and John Day (five children) and Amelia and William Butcher (six children). Another 30 or so mill workers were single men, and perhaps another ten resided in the married men's quarters while their families lived in nearby townships.

The number of pupils attending the No.1 Mill school increased over time, providing another indication of community size. The earliest petitioning for a school in 1904/5 emphasized the 13 resident children of school age. This had increased to 16 by 1908, to 25 by 1916, and by 1923 there were 38 pupils squashed into the small schoolroom (VPRS 795/2812/3601, 23 Feb. 1923). Teachers' correspondence appears to indicate not that family sizes were increasing, but that more workers with families were based at the mill, attracted in part by the provision of a school for their children. This is supported by W. R. Henry's correspondence with the Forests Commission in 1923, when he revealed he had 'about 20 families' on the payroll (FCV 29/0474 10 Aug. 1923). Although it is uncertain whether Henry was replacing single men with married workers, it is likely that the mill population had simply grown since its opening in 1904. The extra men needed to work the No.3 Mill and the reopened Noonday Mill may have lived at the No.1 site as well, and sent their children to the mill school.

Families

In 1870 the average family in Australia included seven children. By 1920 this had shrunk to three children (Hicks 1978: 157; Summers 1994: 368). At Henry's Mill, although there were several larger families present, the number of children per family mirrored this broader social pattern. Women at the site, in common with many others throughout Australia, were choosing to control their fertility and limit their pregnancies to a level well below that of their mothers and grandmothers. This was probably due to several related factors, including later marriage and child-bearing, and better education for women. Limiting the size of the family was also linked with a desire to improve living standards, so that each child could be given a good start in life, as well as women gaining greater control over their own lives. Improvements in public health and sanitation from the later nineteenth century onward lowered infant mortality rates, increased the likelihood of children surviving to adulthood, and reduced the need for large families.

The declining popularity of large families by the early twentieth century was also due to the rising cost of supporting them. For many working families the decision to have fewer children may have been a matter of basic survival, of food, shelter, and future job prospects (Reiger 1985: 119). Following sustained economic growth between 1860 and the 1880s, the effects of the 1890s depression dramatically reduced the economic stimulus of immigration and foreign capital. The first four decades of the twentieth century were thus marked by periods of significant stagnation or decline in real incomes and living standards for working Australians. This was only partially offset by shorter working hours, earlier retirement and an increase in life expectancy. The cost of living during the First World War, for example, soared due to the drought of 1914, the disruption of trade caused by the war, and the comandeering of wheat and meat for military supplies. The basic wage rose by 39% between 1914 and 1920, but the cost of living increased by 72%, leaving families with larger bills but less means to pay them (Boehm 1993: 322-6; Report 1920-21: 58-9).

Grace Karskens has challenged the view that only middle-class women in nineteenth-century Australia were interested in improving the quality of their family life. She cites archaeological evidence from the Rocks in the form of moralising china and hundreds of toys to argue that working-class families cared for and indulged their children. She also uses a bone contraceptive(?) pessary from the Rocks, and three from the 'Little Lon' site in Melbourne, to suggest that women in urban, working-class communities chose to control their fertility rather than risk injury, infection and death through repeated pregnancies (Karskens 1999: 173-9). A similar plastic ring from House B at Henry's, if interpreted as a contraceptive, supports the view that rural women also had access to birth control. Kerreen Reiger has argued that contraceptive knowledge was limited in late-nineteenth-century Australia, but spread rapidly thereafter. She identifies several sources for this extension, including the illicit trade of abortionists, the work of birth control propagandists, newspaper and magazine advertisements, and proselytizing among women themselves (Reiger 1985: 114). Sheaths, pessaries, pills and vaginal sponges were readily available, and in some cases could be made at home from other domestic materials (Quiggan 1988: 105-110). Grimshaw and others have suggested, however, that the decline in conception in this period was effected as much by sexual abstention as by 'the new mechanical devices' (Grimshaw et al 1994: 195). Verandah sleep-outs, noted in the previous chapter, provided an architectural accommodation to this marital arrangement.

The social transformation represented by this shift from large families to small was sharply at odds with the middle-class views of politicians, doctors and church men, who promoted the 'family', with virtuous wives at the centre, as the foundation of society. They regarded motherhood as women's primary vocation. Women controlling their fertility and having fewer children undermined this ideal, and was held to threaten the vitality of the nation and race (Report 1907: 10; Summers 1994: 368-9).

The shift also brought in new ways of seeing marriage, sexuality, housewifery and child-rearing. It strengthened women's role in marriage, by reducing the endless round of pregnancy and breast-feeding, and permitted them to value their sexuality more highly. New standards of hygiene, however, made housework and childcare more demanding and time-consuming. Germs were declared to be a housewife's invisible enemy, lurking in milk, dirt and dust. Housework for rural women, including those at Henry's Mill, became more difficult than for many of their suburban counterparts, because they lacked the gas, electricity, modern appliances, adequate space, sewerage, ventilation and hot running water needed to wage war against the summer dust, winter mud and the dangers they were held to represent. The greater investment in each child reinforced the decision to have fewer. With husbands at work and children at school, women were left to shoulder the burden of maintaining the domestic sphere largely on their own.

It has been frequently noted that employers by this period wanted a disciplined, sober and efficient workforce to support a newly industrializing society (Lake 1986: 130; Shackel 1993: 130-5). Mill owners often preferred to employ married workers with families, even though it meant providing more facilities, because it was believed such workers were more stable and committed to their employment, and were more reluctant to leave in search of better employment opportunities. W. R. Henry had to invest considerable money and effort in providing houses for workers' families and a school for their children, to ensure that his married employees would remain working at the mill. Harold Grenness, teacher at the state school in Forrest, had observed in 1905 that 'The sawmillers [i.e. W. R. Henry & partners] cannot get married people with families, *whom they wish to employ*, to settle in the locality, because there is no school available' (VPRS 795/1927/3601, 9 Jan. 1905, emphasis added). Delays in opening the school and its initial part-time operation in 1909 made it difficult for the firm to retain married workers. In one instance, Henry complained to the Secretary for Education that he had

> just now got them [a man, his wife and four children] settled back again & built houses & am now faced with the same trouble as several of the parents say they will now move off & look for work where they can get a school (VPRS 795/1927/3601, 2 Oct. 1909).

Henry was willing to lobby bureaucrats, exert pressure on the local member of Parliament with the threat of closure and job losses, and threatened to hand 'the whole business over to the press' if the school's operation was not re-arranged to his satisfaction (VPRS 795/1927/3601, 15 Nov. 1909). He got his way a few days later, and the children began receiving full-time instruction. The correspondence demonstrates his commitment to retaining married men by providing a school for their children, and the power of workers who were able to achieve conditions which suited them.

W. R. Henry was not alone among Victorian sawmillers in providing facilities to encourage married workers with families to settle at isolated mills. David McClelland, for example, operated a sawmill at Mount Samaria, near Mansfield, between 1921 and 1923. He built houses for the 14 men and their families, and installed a piped sewerage system to maintain the comfort and hygiene of the 'village'. A school for the mill children and a house for the teacher were also constructed, along with a telephone line to establish communication with the outside world (Evans 1997: 77). McClelland, however, was an engineer by profession, not a sawmiller. He failed to appreciate the relatively poor quality forest in the area, and so the mill closed after only a few years. At the Comet Mill, 50 kilometres north of Melbourne, the Australian Seasoned Timber Company provided a school (S.S. 2799) and neat, high quality housing for families compared to that usually erected at sawmills (Evans 1993: 40; Harvey 1992: 146). The John Hay & Co. No.2 Mill operated at Wye River, on Victoria's west coast, between 1920 and 1921. Half-owned by the Henry Jones-IXL group, the mill was expensively fitted out with houses for the manager and secretary, along with an office, store, boarding house, bakery, school, smithy, stables and 24 houses and huts for employees. The mill's relative isolation necessitated significant expenditure on such amenities, but over-capitalisation quickly resulted in the operation's demise (Scales 1983). It was thus a delicate balance for proprietors to provide sufficient facilities for workers, but at a cost which did not jeopardise profitability.

A similar attitude has been documented in logging settlements elsewhere, especially in North America, where timber companies were also interested in reliable labour. Prior to World War I, demand for work generally exceeded the supply of jobs, so mill bosses had little incentive to improve the generally deplorable living conditions. Workers responded by hopping from camp to camp and indulging in riotous drinking binges (Radforth 1987: 88-106). This pattern of worker transience, however, was disrupted by technological changes which placed an increasing premium on efficient, stable and professional labour, along with the rise of militant industrial unionism. Timber companies responded by repressing union agitation, reforming production methods and improving living conditions. In highland West Virginia, for example, the mixed-gender family logging camp of the early twentieth century proved a successful adaptation by both the companies and families involved. The companies had a reliable pool of labour, families added stability to the traditionally high turnover of workers, and women brought a modifying influence to the often harsh conditions of the logging environment (Brashler 1991). In coastal Oregon, Washington and British Columbia the industry was initially dominated by small producers employing a workforce which was overwhelmingly unmarried and of foreign birth. Labour turnover was around 500 per cent annually (Rajala 1989: 171). Companies established permanent or semi-permanent logging communities in the early twentieth century, in response to high labour turnover and associated costs. The creation of new 'industrial villages' which housed whole families aimed to establish a more stable, docile, dependent workforce than the all-male logging camps of the nineteenth century (Rajala 1989: 173).

Migrations

Following the introduction of the Commonwealth *Immigration Restriction Act* in 1901, the number of non-Europeans living in Victoria declined rapidly. The effort to keep Australia 'white' was very effective during this period, with the numbers of those born, for example, in China, India and 'Syria' declining significantly between 1901 and the 1930s. Non-Europeans formed small, generally marginalised groups in Victoria during these years, where they were denied naturalisation, the vote and the old-age pension (Broome 1984: 129).

The ethnic composition of North American lumber mills during this period was highly dependent on recent immigration from Germany, Ireland, Scandinavia and Eastern Europe (Franzen 1992: 83-5; Radforth 1987: 32-4). In contrast, the Victorian sawmilling industry of the late nineteenth and early twentieth century relied upon workers from predominantly Australian-born, Anglo-Celtic or northern European origins. Analysis of names on Electoral Rolls indicates that the workforce at Henry's was dominated by a similar range of ethnic backgrounds. Gustave and Bernard Seebeck, for example, worked at the No.1 Mill during the 1920s as manager and head sawyer respectively. Although both were born in Victoria, their parents came from northern Germany (Seebeck 1979). W. R. Henry was born in Ballarat of English parents. 'Charlie Han' is the only individual of non-European background documented at the No.1 Mill, probably of Asian extraction (R. Henry 12.6.1999). The main ethnic group identified with a specific forest-based industry in Victoria were the Italian charcoal burners, the *carbonari*, whose slow, controlled burning of timber in earthen kilns produced the quantities of charcoal needed in blacksmiths' shops (Jack and Cremin 1994: 65).

Internal migration within and between Australian states probably accounted for much of the movement to Victorian sawmills. Most timber workers and labourers in the Forrest area, for example, came from Melbourne, country Victoria, the goldfields and Tasmania (Houghton 1976: 8). Although difficult to quantify, it is known that five families came from Tasmania in 1920 specifically to work at the Wye River sawmill (Scales 1983). In another instance, Jack Shaw was a wood-chop event champion in Tasmania. He quit his job as a foreman with the Van Diemen's Land Company at Stanley, and arrived in Victoria around 1928 with his two sons, to tramp from mill to mill in search of work (Davies 2001a: 115). Tasmanian forest workers were sometimes sought in Victorian mills because they were accustomed to hard work, low pay and poor conditions. Migration of workers between local mills in search of better pay and conditions was also prevalent. A. C. Ilie, district forester for the East Otways, reported in 1925 that

> No difficulty is experienced in obtaining men for the mills. They are constantly coming and going from the Warburton, Beech Forest and Powelltown milling districts (Ilie 1925: 5).

In exploring the destinations of those who left Henry's Mill, it is thus likely that many sought work at other sawmills in the Otways or beyond. It is very difficult to document in detail, however, the patterns of dispersal and relocation involving former employees. Amelia and William Cherry and their two children, Florrie and Nellie, lived at Henry's Mill in 1908-9. They returned for another stint in 1921, and as a family were in local demand for their musical skills, the girls on the piano and William as a violinist (CH 13 Sept. 1922, 9 May 1923). Others were probably content to leave behind the mud, toil and danger of the industry for good. Norm Gardner, for example, worked at Henry's Roadknight and Fun Factory mills in the 1930s. After spending a couple of years crosscutting logs, crouched in mud up to his waist, he gladly left the industry to labour on local shire roads (N. Gardner 15 Nov. 1998).

Teachers

Otway Saw Mills School No.3601 was a single teacher school. Fifteen sucessive Head Teachers taught there between 1909 and 1927. All were male. Many were young, and generally had limited teaching qualifications and experience. Nevertheless, the teacher was an important figure in the lives of children at Henry's Mill. Arriving from the world beyond the forest, he represented the authority of the state at the mill settlement.

Initially, female teachers were explicitly excluded from appointment to the school. An early District Inspector had noted the lack of accommodation suitable for a female teacher. Given the isolation of the forest setting it may also have been regarded as too remote for female teachers to cope with. Over time, male-only appointments became an established practice at the mill. However, single young women routinely worked in nearby towns such as Forrest and Barramunga, and at other mill schools at least as remote as Henry's Mill. Blanche Murphy, for example, applied successfully for a position at the isolated Rubicon sawmill school in the forests of the Central Highlands. Her time there, however, was marred by the climate, poor accommodation, lack of companionship, cost of living and bad language of some of the men. Their drunken pranks included hiding the water closet and stuffing rags into her underwear hanging on the clothesline (VPRS 640/P1/1671/4031, 4 Feb. 1921). Alice Hartley was the only teacher posted to the Loch Fyne Mine school near Matlock, in the Central Highlands, from 1899-1902, an extremely remote and rugged location (Carroll 1988: 90). The reluctance to appoint women to the school at Henry's is also anomalous because with so many tiny rural schools opening and closing in this era, finding teachers to staff them was a constant challenge for the Education Department. Forest-based mill settlements were not necessarily any more isolated than many other rural settlements at some distance from towns, railways and roads.

Nevertheless, teachers complained, loudly and often, about the remote location of Henry's Mill. Clifford Stanford, for example, complained in 1911 that the mill 'is [in] an out-of-the-way place, and is far from being a pleasant place to live' (VPRS 640/2328/3601, 28 Nov. 1911). Bernard Flood, in applying for transfer, felt that

> For months at a time, a teacher is practically a prisoner in the gully, for he cannot get away, even on weekends, on account of the remoteness of the district (VPRS 640/2423/3601, 27 Oct. 1914).

A district School Inspector noted the 'wet climate, 6 months practically of very wet weather, and the mountainous character of the country' (VPRS 795/1927/3601, 9 Jan. 1905). Another recorded that the school was 'situated in an inaccessible part of the forest. There is no road to it; no horse can get over the mountains. The only way in is from Forrest by a saw mill tram' (VPRS 640/2423/3601, 1 May 1915). The school's first teacher, Leo O'Kelly, argued for a full-time position because the mill lay 'in a solitude entirely cut off from Forrest and Barramunga' (VPRS 795/1927/3601, 23 Oct. 1909).

Teachers were constantly asking for transfers. The average length of appointment at Henry's Mill was 15 months, the shortest being seven weeks, and the longest almost three years, by Clifford Stanford (1910-1912) and Charles Branditt (1919-1922). Teachers nearly always requested removal from the school, rather than pay rises or allowances to compensate for the isolation. In several cases teachers simply ceased duty without permission or explanation, unable to tolerate the delay in appointing a replacement. Several were dissatisfied with the board and accommodation available at the mill, which involved either the hut in the schoolyard or boarding privately with one of the mill families. William Morris, Head Teacher for a short stint in 1912 and 1913, was reduced to sleeping in a bed with one of his students. Morris later blamed the damp climate for his contracting influenza and neuralgia, and as grounds for transfer to a warmer, inland place such as Ballarat (VPRS 640/2328/3601, 16 June 1913).

Bernard Flood applied for transfer in 1914 on account of the extreme isolation of the mill settlement and the impossibility of doing further study, along with the poor accommodation, damp climate and its effect on his health. Several teachers were also intent on marriage and were reluctant to bring their brides to a school without a proper dwelling. This was a concern echoed by soldier settlers on isolated farming blocks in the 1920s and 1930s, conscious that a tent or humpy was not a fit dwelling for a woman (Lake 1987: 155). Cecil Wallis complained of the financial burden imposed by providing separately for his wife and child at Geelong, and himself at the mill. The pursuit or attainment of further teaching qualifications, especially the 'Second Class Certificate', was also cited as reason for transfer to another teaching post.

Charles Branditt was the only teacher who brought his wife with him to the settlement, thus making their time at the mill more rewarding than arduous. Nina Branditt taught sewing to the schoolgirls, thereby earning £40 per annum and significantly improving the couple's financial position. The Brandits also had the energy and initiative to organise social evenings and a fete to raise money for school books (Branditt 1922). Edward O'Connor, on the other hand, was the only teacher at Henry's Mill ever to request transfer to an even remoter school. He applied in 1926 to teach at the Old Federal Mill, beyond Warburton at Starvation Creek, claiming to know the country and seeking a permanent position (VPRS 640/2639/3601, 1 Nov. 1926). It is doubtful, however, whether many of the other teachers at Henry's Mill appreciated the admiration felt for them by Frank Tate, the Secretary for Education. Secure in his Melbourne office, Tate clearly romanticised their toils in remote places. Enduring the 'hardships and lonely life of the pioneers' was probably a fate most teachers hoped to escape as quickly as possible (Blake 1973, vol.1: 339).

Teachers were not well paid in this period either, especially if they were only qualified and experienced to work in smaller rural schools. In 1912, for example, a 'Class VI teacher', controlling a school of up to 35 pupils, earned about 42 shillings a week (£120 per annum), the same rate as the basic wage. After nine years of service, this could increase to a maximum of £200 per annum (Blake 1973, vol.1: 322). In 1918, women teachers' salaries were raised to at least four-fifths of the rate paid to men of the same class, but the 'family obligations' of men continued to be recognised with higher salaries. By 1920, the young male teacher received £3 per week, and his female counterpart £2 8s. Annual increments over a five year period raised this by nearly one pound a week.

This means that after training at the Teachers' College, and with five years of experience, a male Head Teacher in 1920 earned only 6s 6d a week more than the basic wage. He earned significantly less than a head sawyer or millwright, and the same as a sawmill labourer (TW 26.4.1921). For the teachers at Henry's Mill, approximately one quarter of this weekly salary was needed for boarding house meals alone. A correspondent to the *Colac Herald* argued that the 'totally inadequate salaries' offered by the Education Department was the main reason for the difficulty in finding teachers to work in rural districts of Victoria. This resulted in 'an army of temporary or unskilled teachers who are receiving less remuneration than many of the lower paid hands in timber mills of the Otway forest' (CH 14 June 1920). The difficulty the Education Department sometimes had in filling the vacancy at the Henry's Mill school may thus have been a systemic problem relating to unattractive pay scales, as much as to the isolation and poor living conditions.

Several incidents also suggest that the mill school sometimes attracted poor quality teachers. Edward Prendergast, for example, accidentally locked seven-year-old William Wilkinson in the schoolhouse after hours. With the river swollen by winter rains, concern over the boy's disappearance abated only when he appeared, hours later and injured, having crawled from a window. Prendergast was severely reprimanded by the Department for his 'gross carelessness' and negligence (VPRS 796/761/3601, 3 Aug. 1916). Clifford Stanford was required to remedy defects in his teaching, while Francis Shine was judged by the District Inspector as 'inexperienced and ...doing fair work only' (VPRS 640/2423/3601, 17 Dec. 1916). Bernard Flood was rebuked for his extreme action in suspending Gladys Butcher for disobedience (see below), and warned to 'use sounder judgment in future in the maintenance of [school] discipline' (VPRS 640/2423/3601, 31 May 1915).

Although complaints about the isolation of the mill had lessened by the 1920s, the Education Department found it increasingly difficult to attract applicants when the teaching post at Otway Saw Mills fell vacant. It acknowledged the difficulty in keeping a teacher at the school during the winter months, and there were several periods when the school was closed mid-year for weeks at a time while a replacement teacher was sought. In mid-1926 no schooling was provided for more than two months while a temporary teacher was sought to fill the vacancy. The position was advertised at least 11 times before attracting any applicant (VPRS 640/2639/3601, 9 June 1926). A similar situation occurred the following year, prompting criticism from the School Committee at the reluctance of young teachers to leave the city and thereby deprive 'Country Children of their right to a decent schooling' (VPRS 640/2737/3601, 7 March 1927). The frequent delays in appointing replacement teachers clearly indicate that the mill school was not a highly sought after post, and that most teachers avoided it if possible.

Although bureaucrats in the Education Department were often unwilling or unable to grant the requests of teachers for transfer from the mill school, they were prepared to support teachers in the matter of appropriate accommodation. On at least two occasions, in 1912 and 1915, the school was threatened with closure if suitable arrangements for teacher lodgings were not made (VPRS 640/2328/3601, 9 Oct. 1912; 640/2423/3601, 20 Sept. 1915). Correspondent for the School Committee, Cecil Turner, was then forced to reassure the Education Department of the adequacy of the available accommodation. The erection of a more substantial dwelling was considered in 1916 by the Department, but deferred due to financial stringency, hinting that the economic burden of the First World War on public spending had a tangible impact at Henry's Mill (VPRS 796/761/3601, 9 March 1916).

In many respects the Otway Saw Mills school was typical of the thousands of one-teacher bush schools dotted around Australia by the early twentieth century. The letters of complaint and excuse penned by its teachers have much in common with the fears and anguish expressed by numerous other teachers appointed to similarly remote settlements (e.g. Nelson 1988). The 'bush schoolies' trod a common path in leaving home to attend a training institution, before venturing to another community, entrusted with the task of training the young. Many felt trapped in the isolated districts to which they were sent, cut off from family and friends, unable to cash a cheque or visit a shop for weeks at a time. In struggling to adapt to life in tiny, remote settlements and teach a highly regulated curriculum, their efforts provided the basis of an education for many thousands of young Australians in this period.

Children

The importance of the mill school and its teachers was a reflection of and reaction to the numerous children at the settlement and their need to be raised, nurtured and educated. Children formed a substantial part of the population of Henry's Mill, and in their constant movement and presence were a prominent part of family and social life. The daily activities of mill children were largely dictated by the twin demands of schooling and domestic chores. Gender divisions were clear-cut. Gathering and chopping wood for the open fires and kitchen ovens was a constant job that occupied most boys before and after school. It also provided many with their first contact with axes and saws. Hauling water from the river and tending vegetable gardens also took up time and energy. The domestic duties of girls started early in life, caring for younger siblings and helping their mothers prepare meals, wash dishes and launder and iron clothes. There were also chickens to feed and eggs to collect, lamps to clean and verandahs to sweep.

Under the *Amending Education Act* of 1910, every child in Victoria was required to attend school for the full period of eight years between the ages of six and 14. They were exposed to a range of subjects during their primary school years. These included English (spelling, reading and grammar), arithmetic and algebra, domestic science, history and civics, geography, drawing, music and penmanship. However, the syllabus was formal and restrictive, with rote learning often coming at the expense of understanding. The monitor system ensured that older pupils helped to instruct the younger ones. When girls did needlework inside, boys worked outside in the school garden. Primary school education was completed with attainment of the Grade VIII Merit Certificate. It remains unclear, however, whether many children at Henry's Mill completed their schooling. Disruptions were common, and included family moves, absent teachers and domestic chores. On the other hand, Lillian Brown was awarded a certificate of meritorious attendance for attending school every day for the years 1909, 1910 and 1911 (VPRS 640/2328/3601, 26 Dec. 1911).

Educational opportunities for children at Henry's Mill beyond this level were limited, as the nearest High School was at Colac. Older boys may have worked in the sawmill, with industry awards specifying wage levels for 14 and 15-year-olds (TW 26 April 1921). On occasion, several mothers also sent their under-age three and four-year-olds to the school as well. This prompted the teacher to protest that parents were avoiding their responsibilities and that the school was turning into a nursery as a result (VPRS 640/2423/3601, 15 April 1916). For the children concerned, much of their schoolroom experience was characterised by crowded and cramped conditions, as a single male teacher attempted to convey the basic curriculum to children of varying ages, aptitudes and attitudes.

When school and chores were dealt with, a wide range of games were played. 'Kick the tin', for example, a derivative of hide-and-seek using an empty tin can, was popular. Throwing stones at trees, foot races and games of rounders were enjoyed as well. Building cubby houses in the bush was another pastime. Children at the mill also exploited the play opportunities of the industrial world. They rode on the timber tramway and swam in the pool on the river dammed by sawdust from the mill. Sheets of green bark from the log-yard could be used sappy-side down as toboggans to slide down the sawdust heap (Evans 1994: 112). Play in forest areas around the mill also added food to the pot. Fish were common in the mountain streams, including the grayling, short-finned eel and blackfish, along with the introduced brown and rainbow trout. The children were eager fishers and enjoyed spearing, hooking or netting the fish. They also trained ferrets to catch rabbits, and collected blackberries when in season. Playing around the mill, however, could also be dangerous. Thomas McGlone, a school pupil, was severely burned when he fell onto the burning sawdust heap next to the mill shed (BT 23 July 1918).

Archaeological analysis also identified the remains of more than 40 toys and other child-specific items. These included china and plastic dolls, marbles, tea sets and gaming pieces. In addition, there were at least 30 slate pencils. Highly valued items were presumably well cared for and taken away when a family left the mill. Although the archaeological record does not express the full range of material culture used by children, the remains demonstrate the diverse and prominent nature of children's items at the site.

Some artefacts, including German-made bisque doll heads and Bennington marbles, were of exceptional quality and were relatively expensive. Parents of at least some children were willing and able to purchase high quality toys for their children, in spite of the award-level wages paid to sawmill workers. Alternatively, these may have come from grandparents or other relatives, or were handed down from earlier generations. Many of the toys were also very contemporary. In spite of their physical remoteness, children at Henry's Mill were able to acquire up-to-date, fashionable playthings. Living on the geographical margins did not necessarily mean that children were deprived of the latest toy offerings. Access to the outside world by mill children is also indicated by Ric Henry's memory of walking with friends along the tramline into Forrest township and back again to enjoy the novelty of an ice cream in summer. The mill's isolation could be overcome by children as well as by adults, when the time and inclination to walk 20 kilometres were available.

Children also made their own toys. The industrial and domestic environments provided a wide array of potential play objects, from washers and clothespins to bottle tops and pieces of string, adapted or renamed in the course of play. Children modified materials familiar and available to them. Fishing nets were crafted from old pieces of flywire to scoop up minnows from the river. Bows and arrows were made from the local satinwood tree and used for games of Robin Hood. A stick could be almost anything, becoming a horse to ride, a log for building, a boat to sail or a gun to shoot. Girls skipped rope and played house with rag dolls. In fact, it was these toys the children made themselves whose loss was felt most keenly when the mill burnt out in 1927 (R. Henry 12 June 1999).

Children were thus active and creative manipulators of their social and material environment. As well as crafting their own playthings, they could also acquire mass-produced toys through barter with peers, purchase with pocket money, and theft. They could ignore, lose or destroy toys that did not suit them, and use them in ways never intended or imagined by the manufacturer. Toys and other children's items were a vital medium of symbolic communication between both adults and children and among children themselves, as they negotiated identity and status within their peer groups.

Individual People

Having examined general patterns of population, family and migration at Henry's, and considered the experiences of teachers and children, the following section presents brief accounts of several individuals and families at Henry's Mill. The people documented are those whose prominence, adversity or reputation at the site generated enduring memories and/or historical records. The aim below is to engage with the working lives and experiences of people who comprise a rough cross-section of the mill community. From their stories emerges greater understanding of the fears, hopes and tensions common to the people of Henry's Mill and the wider community beyond.

Frank McGlone

Four McGlone brothers, Thomas, Francis, James and Herbert, worked for the Henrys as mill hands and fallers in the late 1910s and early 1920s. With their wives and children they formed a sizeable clan within the mill community. The family is noteworthy for its frequent association with petty and serious disputes, including both property offences and alleged physical assaults. The documented illiteracy of Francis and the family's poor attitude to learning suggest that they attached limited significance to education. All four brothers worked in occupations requiring little or no formal schooling.

Francis Lindsay McGlone began working as a faller in the forests around the No.1 Mill in 1917. He was married with four children, and a noted marksman, but illiterate. His brother Thomas also worked at the mill, with his wife Claudia and their seven (later, eight) children. In 1918 Francis McGlone was accused of the sexual assault of Thomas and Claudia's daughter, 14-year-old Vidah McKreena McGlone, on the night of 1 August (BT 22 Oct. 1918). Vidah's family lived at the No.1 Mill, but at the time her parents were away in Colac. Francis McGlone had been asked by Vidah to stay over for two nights to keep her and her siblings company. She later claimed her uncle had spent the first night with her in her bedroom. Under questioning at the Birregurra Police Court, however, Vidah admitted that she had lied in her statement and that her uncle had never touched her. Her mother, Claudia, admitted that she had had 'a good deal of trouble' with Vidah through her daughter's involvement with young men. Vidah herself acknowledged that she had been sexually active since she was about 13, and had accused her uncle out of fear of her father giving her a hiding for 'carrying on with' boys from the nearby township of Barramunga (BT 22 Oct. 1918).

Francis McGlone maintained his innocence through the entire legal process, while the police admitted that all of his statement had been prompted by their questions. His guilt or innocence appears to have hinged on a medical examination of Vidah, and whether he spent the night in his niece's bedroom, or merely sat on the edge of her bed to talk. He was committed to trial at the Geelong Supreme Court three weeks later. On Armistice Day, 11 November 1918, McGlone was acquitted of the charge and walked free (VPRS 3524/P1/26).

While the bare detail of the ordeal is recorded in local newspapers and court records, the emotional strains and tensions within the McGlone family at this time can only be imagined. Although Claudia McGlone had had to deal with her daughter's misconduct before, she had become very upset at the time of the police questioning at the mill. Her husband, Thomas, appears from the newspaper account to have been a man accustomed to responding with violence, in word and deed, at times of crisis. It is unclear whether he believed his daughter or his brother, but his first response was to knock 'their b—heads together' (BT 22 Oct. 1918). The McGlone family also had to face the judgment and disapproval of the mill community on a daily basis. The response of Francis' wife, Elizabeth, and how they coped as a family afterward is not recorded.

This case, however, is only the most dramatic of a range of incidents involving the McGlone family. In 1917 Francis, his brother, Herbert, and another man, Joseph Smith, were accused of assaulting and robbing Meery Aman in Forrest of 30 shillings (VPRS 292/7, 28 Feb. 1917). The following year, soon after the alleged offence against his niece, Francis McGlone was taken to court for failing to pay £3 he owed to John Kirkpatrick for work and labour (VPRS 292/7, 18 Sept. 1918). In 1921 he was fined 40 shillings for driving an unregistered motor vehicle in Birregurra without a license (VPRS 292/7, 12 Oct. 1921). Fourteen-year-old Rupert McGlone was the subject of complaints by his schoolteacher, Cecil Wallis, for irregular attendance, laziness and disruption (VPRS 640/2563/3601, 8 March 1923). Thomas McGlone took his employers, W. R. Henry & Son, to court in 1922 over their alleged failure to pay workers fortnightly. The proprietors had contrived to allow their employees to draw their pay whenever it was convenient, a practice in which the men concurred, even though it contravened the terms of the industry award (BT 4 April 1922). Two-thirds of the men did in fact draw their pay every fortnight. Nevertheless the Police Magistrate was obliged to record a conviction against the Henrys of 40 shillings with 21 shillings costs. The proprietors must have felt frustrated at being prosecuted for obliging their men their wages, perhaps resulting in a frosty relationship with Thomas McGlone thereafter.

The activities of the McGlones in Forrest and Birregurra, and the time spent by Vidah's parents in Colac, reinforces the accessibility of the outside world for mill residents. The disruptions caused by members of the McGlone clan, however, undermined the notion of married workers as more stable and reliable than single men. The conflicts in which they were involved indicate that domestic scandals and crimes were as present at the No.1 Mill as anywhere else. In small rural communities where privacy was hard to maintain, tensions and disputes within and between families were difficult to conceal and contain. The decision to move away and thereby avoid conflict, when money and work skills were limited, may not always have been simple.

Gus and Bernie Seebeck

Gustave and Bernard Seebeck were two of seven children born to Johann and Johanna (née Schmidt) at Chewton on the Victorian goldfields in 1866 and 1871 respectively. Both followed their father in the sawmilling trade, and worked for W. R. Henry & Son in the 1920s. The work, movements and business interests of Gustave especially are an example of the peripatetic lifestyle pursued by a talented and hard-working family man in the late nineteenth and early twentieth century.

In 1895 he packed up his wife, Emily, and their daughter, Francistra, from their home in Echuca and headed to the goldfields of Western Australia. Gustave knew from the beginning that the surest path to success in the industry was not in mining but in supplying provisions. Over the next 12 years he engaged in a number of business partnerships in the region, shifting between Coolgardie, Kanowna, Broad Arrow and Menzies every year or two (Seebeck 1979). He was involved variously with the manufacture of aerated waters and cordials, distilling water, cutting timber and managing sawmills. His repeated moves suggest that these partnerships were not always very successful. During this period Emily Seebeck gave birth to six children and accompanied her husband on his frequent moves.

In 1907 the family returned to Victoria, and Gustave began sawmilling in the Macedon Ranges, before moving to the Warburton area. By the beginning of World War I the family had settled in Northcote, Melbourne, by which time Emily Seebeck had had another four children. Gustave appears to have spent some time cutting huon pine around Strahan in western Tasmania, before working with the Henrys as mill manager from 1922 to 1929. It is not clear whether his large family accompanied him on these later moves. During this period, Bernard Seebeck also worked at Henry's Mill, as head sawyer. Remembered as 'a tough old bird' who would saw timber all day in his bare feet, his skill with a violin meant he was always in demand for bush dances (Facey 1977: 11-12). Emily Seebeck died in 1925, and eventually Gustave settled in Melbourne to raise poultry, where he too died in 1936 (Seebeck 1979).

It seems likely that the paths of Gustave Seebeck and W. R. Henry crossed during the 1890s in the Kalgoorlie goldfields, where Seebeck maintained timber and merchandising interests and where Henry was a successful mine manager. Many years later, the two forged a working partnership at Henry's No.1 Mill. Paradoxically, it was here that Gustave found some stability in his roving life, remaining with the Henrys for seven years, in an industry characterised by the mobility of its workforce. He may also have felt encouraged to remain at the mill by the continued presence of his younger brother. In other respects his fortunes typify the nomadic life of many bush workers in this period. As head of the household, men had the responsibility of finding work and earning money to support their families. The stagnant Australian economy of the early twentieth century meant that much work, especially rural work, was short-term and irregular. Families frequently followed their menfolk, enduring a life of discomfort and disruption made familiar and more secure by the ties of kin and shared experience.

Gladys Butcher

William and Amelia Butcher arrived at Henry's Mill with their six children early in 1914, and William began work as a mill hand. Later that year, their eldest daughter, Gladys, sparked a dispute at Henry's Mill that provides rare insight into the social conflict that could emerge between individuals and groups at a remote forest sawmill. Two weeks before Christmas, eight-year-old Gladys was marched out onto the floor of the schoolroom by the teacher, Bernard Flood, for insolence and disobedience. She then kicked him in the knee, and was promptly suspended from school. This incident was the catalyst for a long-running dispute at Henry's Mill, involving three successive school teachers, Gladys's parents, Amelia and William, the School Committee, W. R. Henry, the District Inspector and the Education Department. The dispute reveals some of the fears and emotional tensions which could emerge in such a community, the depth of feelings involved, and the social isolation that could result.

The flurry of letters which emerged from the dispute give voice to all the participants except young Gladys Butcher herself. Her suspension continued until March the following year, when the Education Department instructed the teacher to re-admit her. Dissatisfied with the lack of consultation in the affair, William Butcher refused to send his daughter back to school until June, when a meeting of the School Committee negotiated her return to school attendance. There was further dispute several months later when three of the Butcher children (Gladys, Clarice and William) were kept at home by their mother for reasons that are unclear but involved poor behaviour on the children's part and a reaction to the school teacher and mill community by their parents. Another round of complaints against the teacher commenced in December 1916, by which time the authorities involved had wearied of the turmoil, and threatened the children with expulsion unless they and their mother conformed with the discipline of the school.

Although the dispute began with a teacher disciplining a pupil, it soon escalated to one where it brought the parents of Gladys Butcher into conflict with most of the mill community. William Butcher regarded the teacher, Bernard Flood, as 'the boy', while Amelia Butcher later complained that Flood's successor, Edward Prendergast,

> 'is mad...and certainly cannot teach. I think that all the worst teachers in the Department have had a chance here...surely our children deserve a good teacher just for the way of a change' (VPRS 640/2423/3601, 7 Dec. 1916).

Teachers were also subject to verbal abuse by the Butchers, and the threat of physical violence was not an empty one. In 1918 Amelia Butcher was convicted of unlawfully assaulting John Hutchison in Forrest and fined 40 shillings (VPRS 292/7, 26 June 1918). Bernard Flood endured 16 months during which William Butcher 'continually threatened to assault me for reprimanding this child' (VPRS 640/2423/3601, 14 April 1915). In response, teachers and others were certain that the Butchers encouraged the disobedience of their children and prompted their interruption of school discipline. Bernard Flood hastened to assure the Education Department, however, that the Butcher children were 'the only troublesome scholars in the school – the other children never give any trouble' (VPRS 640/2423/3601, 14 April 1915).

The dispute continued to escalate. William Butcher refused to send Gladys back to school, expressing concern to the Education Department about bruising on his daughter's arm, her torn dress, the teacher's rudeness and the lack of consultation in the affair. He argued that Gladys was 'a quick child to learn and can be led but not driven'. W. R. Henry, however, regarded William Butcher's complaints as 'frivolous', and thought the whole family 'headstrong', while the School Committee and District Inspector concluded that the attitude of the parents was quite unreasonable (VPRS 640/2423/3601, 2 June 1915). It was at this point that Amelia Butcher put pen to paper, hinting at the underlying causes of the dispute which had set her family at odds with the rest of the mill community.

She regarded the dispute over her children as a pretext to force her family out of the settlement, convinced that 'they could not make us move any other way but by getting the children turned from the school knowing we would have to go elsewhere'. She believed the situation was

> 'a get-up between the correspondent of the committee & the teacher, The people here have never liked us since we came…they all more or less drink some of the women and all but one or two men my Husband being a strict abstainer neither drinking or smoking [n]or does he gamble We do not mix with them' (VPRS 640/2423/3601, 10 Dec. 1915).

These outbursts reveal her belief that her family had been ostracised by the mill community. She blames this social distance on the participation by most in drinking, gambling and smoking, and her husband's refusal to indulge. In spite of the mutual dislike emerging between members of the Butcher family and others at the mill, however, they appear to have endured the conflicts and remained at the site for at least three years, but the timing of their departure is not recorded.

A strong sense of injustice is evident in Amelia Butcher's complaints, although it is uncertain whether she and her family were merely argumentative by nature, exaggerated the reasons for the dispute or had a genuine grievance. The length of the dispute suggests a reluctance to yield to the social pressure to move away, perhaps due to either personal resilience and resolution, or a lack of employment opportunities elsewhere. There is nevertheless a sense of histrionics, even paranoia, in her letters, dramatised by underlined headings such as 'Only Justice required!', and the observation by the school District Inspector that 'Mrs Butcher appears to be a woman of violent temper' (VPRS 640/2423/3601, 24 Feb. 1916). In taking the high moral ground against drinking and gambling, she conveniently overlooks the inconsistency of her own sexual morality, having borne four children out of wedlock, and emerges as a woman not generally governed by conventional expectations of behaviour.

In the end the District Inspector, frustrated at the attacks and insults of the Butchers, called their bluff and recommended they pursue their grievances in a court of law, knowing there was no substance to their complaints. The Education Department supported its teachers and inspectors, and advised Mrs Butcher that her attitude was unreasonable and vexatious, threatening expulsion of her children unless all submitted to school discipline. Here the matter appears to conclude, at least in documented form. It nevertheless remains unclear how the ill-feeling generated by the dispute was resolved or if it continued to smoulder away in the daily interactions of the participants.

It is unclear whether anybody benefited or 'won' in the Gladys Butcher dispute. The Butcher children, although wilful and disobedient, became pawns in an ongoing argument between their parents, the School Committee and the mill community. Gladys and her siblings had their schooling disrupted and endured unknown taunts from the other children. The school teachers involved felt physically threatened by William Butcher and had their handling of the affair questioned by their superiors. W. R. Henry was forced to intervene as well, chairing meetings of the School Committee and penning numerous letters to try to calm the affair and ensure the smooth running of his business. William and Amelia Butcher, in persisting with the dispute for several years, appear to have been effectively marginalised by the mill community. Their reluctance to yield to social pressure to move away may have owed something to personal resilience or resolution, or a lack of opportunities elsewhere.

Cecil Turner

Cecil Turner was one of the Henrys' longest-serving employees. Born in Colac in 1889, he trained as an apprentice in the Geelong timber yard of Thomas Higgans soon after the turn of the century. He began work at 'Otway Sawmills' a few years later and was still with the firm in the 1930s. As a qualified engine driver, he helped install the first steam logging winch in the Otways around 1908 (Houghton 1975: 62). His resourcefulness was also evident in later years when, with George Hall, he built log lines along the Barwon, Roadknight and Mount Sabine tram routes. Although lacking formal qualifications in surveying, both men had a fine eye for practical gradients in rugged and difficult terrain.

Cecil and his wife Louisa were also long-term residents of the No.1 Mill, along with their children, Jack and Marjory. He served as correspondent of the School Committee from around 1911 until its end in 1927, lobbying the Education Department for desks, blackboards, urinals, spouting and other materials. His argument for a new, replacement schoolroom in 1916, however, was unsuccessful. On several occasions he was also forced to reassure the Education Department that appropriate and adequate accommodation and board *was* available at the No.1 Mill, in spite of protests by teachers to the contrary. He recognised the important role the school played in the life of the mill community, and was prepared to commit time and energy to ensure its survival. At the same time, he had a vested interest in defending what were fairly poor living conditions endured by the various mill teachers.

Turner was renowned as a man who had never ventured north of the Colac-Melbourne railway line (N. Gardner 15 Nov. 1998). He represented an alternative approach to rural living, in an industry characterised by the frequent movement of its workers. It was not lack of opportunity, however, which prevented him from moving on. The forests, farms and townships of the Otway region provided the interest and opportunity he needed to live a modestly prosperous and respected life. He also appears to have been a man of considerable ingenuity, intelligence and energy, adapting available technology to meet the challenges of logging in a remote forest environment. He embodied the rural tradition of improvisation and making-do, earning quiet admiration over the years from workmates as a skilled, experienced bush worker.

Oscar & Eva Henrikson

Oscar 'Stump' Henrikson began working for the Henrys around 1926 as a labourer, while his wife, Eva, ran the boarding house at Henry's No.1 Mill. They worked at several of the firm's mills well into the 1940s, and are strongly and fondly associated in local memory with the cookhouses, or 'Hash Foundry', for which they were responsible (e.g. J. Biddle 2 Nov. 1998; J. Washington 24 May 1999). Norm Gardner remembers Stump Henrikson as 'the hardest worker in the Otways', cultivating large plots of vegetables and keeping a milking cow (N. Gardner 15 Nov. 1998). Eva Henrikson, however, is likely to have been just as hard a worker, as she kept up her role as wife and mother, and prepared dozens of meals for hungry men, six days a week for more than 20 years.

Stump Henrikson was noted for several qualities. He walked with a limp, derived from a leg injured in an accident, while his severed left hand generated his nickname. He was also evidently a fine gunshot, collecting trophies for his efforts at the Melbourne and Geelong Shows. He practised shooting against low-flying black cockatoos, and was never seen to miss. The large garden he kept produced a wide range of vegetables in the rich mountain soils of the Otways, providing abundant fresh food for the mill boarding house (R. Henry 12 June 1999). The Henriksons were clearly a tough, hard-working family, committing the most active part of their working lives to the Henrys' forest sawmills. Oscar was evidently not one to let a damaged leg and missing hand prevent him from manual work with timber and spade, and nor did his wife shirk from many years of domestic toil in remote mountain forest areas. While injuries restricted his work around the mill, Oscar and Eva compensated by running the boarding house and growing vegetables, a strategy which ensured their economic survival.

W. R. Henry

William Robert Henry was born in Ballarat on 12 July 1858, son of Alexander Henry (b. 1829), a miner from Malta, and Emma Johanna Cronk (b. 1837), from the village of Windlesham in County Surrey (Davies 2001: 129). Trained as a plumber, he achieved success in the gold mining industry in Kalgoorlie and Ballarat, before spending the second half of his life as a sawmiller and retail timber merchant (GA 26 Nov. 1957; Houghton 1975: 51-73). Henry's wife, Alice, appears to have spent little time in the forest milling areas, maintaining instead the family home in Geelong. William and Alice Henry had two children, Ida and Ernest. Ern was born in New South Wales in 1885, and trained as chemist in Ballarat before joining his father in the timber business in 1907. He married Gladys Taylor in 1910, and together they had three children, all born in Geelong: Fred (b. 1911), Ivan (b. 1915) and Cedric (Ric, b. 1917). As both William and Ern Henry established their homes and families in Geelong, their primary personal and business attachments lay in town rather than at their forest sawmills.

After returning from Kalgoorlie in 1902, W. R. Henry rapidly assumed financial and operational dominance of the 'Otway Sawmills' business. His willingness to secure finance for the tram tunnel, the first of only three ever built in Victoria, is evidence of a courageous but shrewd operator. The expense was substantial, at over £1000, but the reward justified the risk, as the subsequent operation of the No.1 Mill for the next 24 years was to prove. Familiar with mining machinery, Henry was also willing to adapt technological solutions to the problems of logging in remote forest areas. He installed the first steam logging winch in the Otways around 1908, and replaced horse teams on the access tramline with a steam locomotive in 1911 (Houghton 1995: 6). In 1919 the firm assumed full control of the Geelong timber yard in which they had held an interest since 1909 (Sands and McDougall 1919). In later years, however, Henry appears to have spent less and less time at the No.1 Mill, leaving much of its day to day running in the hands of his son and business partner, Ern. 'Old Man Henry', according to George Facey's memoirs, would ring up from Geelong when he was travelling down to the mill. He would then be met at the Forrest railway station by a driver and a rail car, journey out to the mill, have a look around, and return to Forrest in time to catch the 3.40 p.m. train back to Geelong (Facey 1977: 36).

W. R. Henry & Son suffered various setbacks over the years which undermined their business success. In 1919 the No.2 Mill was destroyed by fire, followed several weeks later by floods which destroyed numerous bridges and tramways (BT 11 March 1919; CH 19 Feb. 1919). The No.1 Mill was destroyed by fire in 1927. Surprisingly, neither of the razed mills was insured, resulting in substantial losses to the firm (CH 31 Oct. 1927). Exposing thousands of pounds of mill machinery to the hazards of floods, accident and fire without insurance appears to have been a high-risk strategy. Given W. R. Henry's success in negotiating finance with banks and royalty arrangements with the Forests Commission, it is likely to have been a deliberate policy, rather than an inability to reach agreement with an insurance company. Nevertheless, the risk did not pay off, and large losses were incurred.

There is also some evidence to suggest that W. R. Henry was a poor judge of forest quality. Following the floods of 1923, the firm invested £1100 in a new tramline to tap timber supplies on the top of the Otway divide. The timber proved to be of very poor quality, however, with many trees shattering so badly upon falling that not a mill log was recovered (FCV 29/0474, 9 Jan. 1926). Between 1932 and 1935 the firm operated the Roadknight Creek Mill, west of the Forrest township. W. R. Henry insisted there was good timber in the area, but his son believed, correctly, that many of the trees were infested with white ants, and the mill did not pay (R. Henry 12 June 1999). It may be significant that the elder Henry was originally brought into the operation on account of his management and financial experience, rather than for his timber-getting knowledge.

Torrential winter rainfall in 1923 washed away 6.5 kilometres of tramline, including about 1500 metres of bridges and makeups (FCV 29/0474, 4 Aug. 1923). The damage bill of around £2000 was exacerbated by keeping around 20 families and a few single men on the payroll, at a cost of over £100 per week. His reluctance to dismiss them suggests both sympathy with their position and a keen self-interest. Henry was aware of the dependence of the men and their families on him, and concerned at the difficulty of gathering a large body of workers later on should he have to dismiss them. Having gone to the trouble of providing housing and a school, he was willing to absorb financial loss to ensure married workers and their families remained at the mill. He argued that it was against his nature to discharge his workers, and felt that it would 'make an excellent whip for the [Timber Workers'] Union to flog us with' (FCV 29/0474, 10 Aug. 1923). He was not above using this philanthropy, however, as leverage to negotiate a reduced royalty, threatening to discharge employees if he did not receive a favourable response from the Forests Commission.

By the 1930s and 1940s, W. R. Henry was recognised as an authority on all matters pertaining to the timber industry, and was frequently asked to speak on behalf of the sawmillers' trade before the Arbitration Court. Although formally retiring from the business in 1934, he maintained an active involvement in the firm until 1938-9, when he helped to install the Nettle and Carisbrook mills (Houghton 1975: 71). W. R. Henry died on 25 January 1949 in Geelong, aged 90, a respected and modestly wealthy man. In spite of the many vagaries of a long life in the mining and milling industries, his estate indicates the success he achieved in his business and professional life. He left his heirs a substantial ten room brick residence in Geelong, 700 acres of land in the vicinity of Forrest, and a timber yard in Shepparton (VPRS 28/P3/4930, 415/726).

A veiled portrait of W. R. Henry as a businessman and employer can thus be built up from various sources, including government correspondence, newspaper reports and oral accounts. He emerges both as a sympathetic employer and innovative businessman, self-made, willing to take risks, and disinclined to accept the authority of bureaucratic, legal and trades union officers. He enjoyed both success and failure in the timber industry, inevitably so for one directly involved in it for around 40 years. Substantial losses came from losing uninsured mill plant to fire and flood, and an inability to judge the quality of timber consistently. Ric Henry recalls that his grandfather 'made two fortunes and went through both of them'. The first is obscure, probably involving gold in Western Australia. The second concerned a major interest in a Melbourne iron foundry. Selling out prior to World War II meant that he failed to capitalise on the firm's monopoly of manufacturing tracks for military tanks (R.Henry 12 June 1999).

With the gloss of years, former workers generally recall the Henrys as good people to work for. Jim Biddle, who worked at the Carisbrook Mill in its final years during the late 1940s, recalls that the Henrys were 'terrific people' who looked after their men (J. Biddle 2 Nov. 1998). Jim Washington also remembers 'Father Henry' as good to work for (J. Washington 24 May 1999). George Facey reiterates this view, observing that all of W. R. Henry's employees and associates regarded him as 'a grand old man' (Facey 1977: 11). Elsie Chamberlain thought him 'a wonderful old gentleman', and enjoyed a weekly afternoon tea with Henry at the Carisbrook Mill in the late 1930s (E. Chamberlain 7 June 1999). Joyce Harding is one of the few who recall W. R. Henry with something less than affection. She and her husband Leo arrived at the Nettle Mill in 1939 as newly-weds, spending the next two years at the site. She recalls Henry, in his early eighties by this stage, as a 'rather sour' man who scarcely spoke to her (J. Harding 21 May 1999). It may have been the 60 year age difference in this case, or simply that Henry was tiring of the many long years commuting between Geelong and the Otways and camping in a hut.

Conclusion

Longer-term workers at Henry's Mill, such as the Henriksons and Turners, enjoyed the benefits and endured the discomforts of life at the mill settlement. They made the best of the living conditions and maintained a distinct affinity for life and work in an isolated bush environment. Others felt very differently, staying only a few weeks or months while they saved a few pay cheques and left the mill at the earliest opportunity. Most people, however, were able to cope for a couple of years at least, many raising families in the hope that their children would be better educated and enjoy greater freedom of choice than their parents before them.

For many women, transition to life at the mill settlement may often have been an isolating and lonely experience. Deprived of the comfort and company of family and friends, they had to cope with basic living conditions and the absence of many household amenities largely taken for granted in urban areas. Domestic and child-rearing responsibilities severely restricted their opportunities to move far beyond the mill. Many left the forest to visit towns only two or three times a year. The ability to adapt quickly and make friends was thus an important asset for a wife and mother in such circumstances. Having young children ensured common interests and concerns between mothers, easing the social absorption into the mill community. It was probably easier for the men, who had an instant peer group with their workmates, among whom they might find friends known from other mills.

In spite of the often uncomfortable and difficult conditions at the No.1 Mill, life and work there had its compensations. The Henrys, for example, were by most accounts good bosses to work for, and they paid salaries a margin above the basic wage. Accommodation was cheap, even if houses and huts were small and of poor quality, and probably little different to those of other rural people. Work in the clean cool air of the mountain forests was steady, although also arduous and frequently dangerous. The No.1 Mill provided basic amenities, including a school, boarding house and post office, but it lacked such refinements as sewerage, running water, electricity and gas. Within the broader context of rural life, working 'in the timber' was a respectable occupation.

Disputes and crimes involving such families as the Butchers and McGlones, however, demonstrate how easily conflict and antagonism could erupt and persist in small communities. People of limited means and education had to fend and speak for themselves, relying on the support of family and friends when injured, wronged or marginalised. These conflicts also show the willingness of some to make a point within their community with verbal abuse and physical violence. The following chapter considers this notion of community in greater detail, examining both the pressures undermining and attitudes reinforcing a sense of communal belonging at isolated sawmill settlements.

Chapter 7

Community

Not everybody felt like they belonged at Henry's Mill. Some felt alienated by the climate and hard work, by their religion or politics, or by their lack of social connection to others at the site. Many regarded the mill simply as a place to work for a while before moving on to other things. Others, however, felt a greater sense of attachment. Cecil Turner, for example, who worked for the Henrys as an engine driver for many years, also headed the School Committee. John Brown worked as a timber trucker for more than 20 years before he was tragically killed on the job. For others, attachment to place was expressed by participation in sports days and picnics, and by the cultivation of gardens and exotic trees. This chapter traces the ways in which a sense of belonging and attachment was evident at the mill, and the cross-cutting loyalties and contested values that divided individuals and groups. It considers some of the ways in which personal differences were expressed, including industrial relations and sectarian divisions. It also looks at some of the positive manifestations of community at Henry's Mill, including school fetes and gardens.

In this study, 'community' refers to a group of people interacting daily in a specic place. An important element of community is the degree of regular, personal interaction, and the ways this is facilitated by social institutions, the workplace, residential proximity, and kinship and friendship. Conrad Arensberg has noted that communities are as often fragmented into estranged groups, factions and sects as they are harmonious or concerted (Arensberg 1961: 247). In Melbourne's 'Little Lon' precinct, Mayne and Lawrence argue that neighbours' sense of place was conditioned by webs of friendship and animosity, of collaboration and exclusion, mediated by gender, social status, ethnicity, race and age (Mayne and Lawrence 1998: 96). The community also shares a common economic base, a sense of attachment to place and shared cultural values. It involves a willingness to participate in communal life, and the security that flows from it.

Residence and mobility

A sense of identity between individuals, a feeling of community spirit, is enhanced by frequent face-to-face interaction over a long period of time. At 'Smalltown', Ken Dempsey's sociological study of a rural agricultural community in north-west Victoria, the length and continuity of residence helped generate an emotional bond that blurred distinctions of class, religion and age (Dempsey 1990: 91-2). More than half of Smalltown residents had lived there for at least 30 years, and three-quarters for at least ten years. Many people had grown up in Smalltown, as had their parents and grandparents before them, and they enjoyed a greater sense of belonging from this generational attachment to place. Family ties and friendship reinforced feelings of belonging and the degree to which community members know one another.

At forest sawmill camps, however, these bonds were harder to forge and more easily broken, as people came and left on a regular basis. Norm Houghton has described the floating population of skilled and unskilled men who laboured in the Otway forests in the early twentieth century. Skilled hands could adapt to the conditions, but others found the industry far from their liking. Many would-be timber workers arrived in the Otways, worked a bare fortnight, and quickly departed. Urban dwellers from Geelong or Melbourne often approached the large timber companies in search of work, and were sent to these firms' Otway sawmills. Rather than the easy work they had anticipated, a few days of dirty, hard and dangerous toil in the wilds of the Otways quickly imposed the reality of bush work upon them, and most departed with their first and only pay cheque (Houghton 1992: 84).

At Henry's Mill, there was a similar pattern of mobility. Evidence from Electoral Rolls indicates that only nine per cent of residents lived at the site for more than five years, and four per cent for more than ten years. In contrast, 84% of residents remained at the mill for less than five years. The shortest documented stay at the No.1 Mill was by two fallers from the Mississippi Mills near Warburton who, in W. R. Henry's words, 'would only stay a little over a week as they said it was too rough for them' (FCV 29/0474, 1 June 1921). A teacher, Arthur Woods, began duty at the school on 28 April 1919, but resigned only seven weeks later, citing health problems brought on by the climate of the area (VPRS 640/2498/3601, 16 June 1919). In the eyes of the mill proprietors, in fact, their employees were always coming and going (TW 31 Aug. 1925). In addition, there are only nine documented examples of family members working together at the mill during its 24 years of operation. Friendships formed in the rugged conditions of life and work at the mill were continually disrupted by the departure of individuals and families in pursuit of better opportunities elsewhere. Such mobility was a major obstacle to the creation of a locally based forest community. At the same time, however, this constant movement *between* mills increased the likelihood

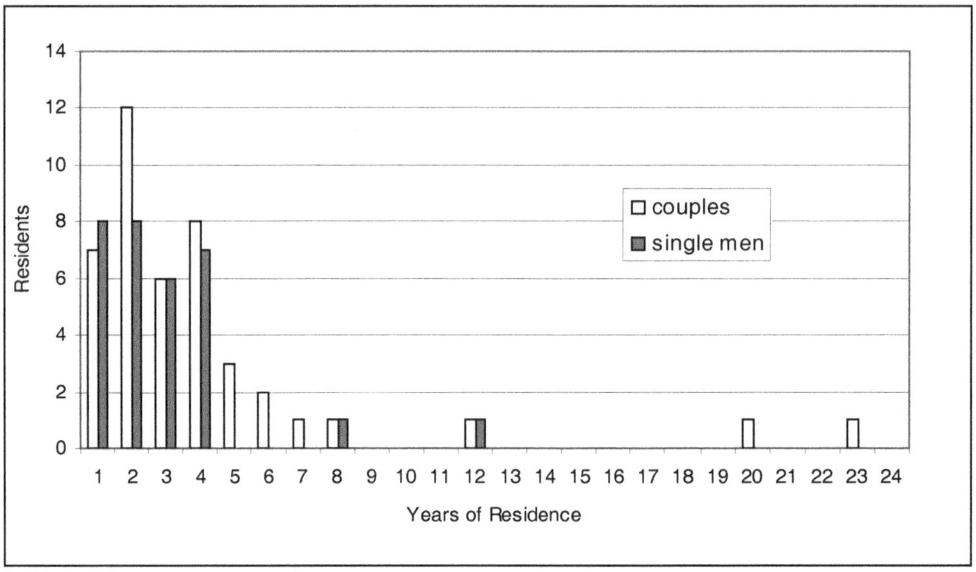

Figure 14: Residential longevity for married and unmarried workers (data source: Victorian Electoral Rolls 1905-1930).

of renewing old friendships, thereby easing the transition into a new sawmill community.

Evidence from Victorian Electoral Rolls also reveals that there were no significant differences in the length of time families or single men spent at the mill (Figure 14). In spite of the Henrys providing a school and housing to attract married workers with resident families, their best efforts appear to have made little difference over the longer term. Both single men and families tended to remain at the mill for no more than about four years. Those who resided longer than this generally worked in more senior, better paid positions, including foremen, millwrights, head sawyers, blacksmiths and engine drivers. The decision to remain at the mill or to move on would have been made with reference to many factors, including employable skills and experience, future aspirations for self and family, satisfaction or otherwise with wages and living conditions, job opportunities elsewhere, and personal ties and relationships with other people at the site. The mill's workforce and residential population was thus turning over every three or four years. A common pattern was for workers and families to arrive, set up their lives at the mill, and typically leave again a few years later. Three or four years of life at the mill, however, would have been long enough for most people to have called it 'home' and felt some sense of belonging within the community.

Politics

Henry's No.1 Mill served as an official polling booth for the electoral division of Corangamite. It was one of seven polling stations in the area, due to its size and geographical position. Electroal Rolls had been prepared annually for the Commonwealth government since 1901, with women gaining the right to vote in federal elections in 1903 and in Victorian state elections in 1908 (Summers 1994: 393). Voting was not compulsory, however, until 1924, and only those 21 years of age or more were eligible. Voting results for three Victorian state elections, in 1920, 1921 and 1924 were recorded in the *Colac Herald*. These results shed some light on voting patterns at the mill and the different political allegiances of its workers and residents.

1920

J. Black (Farmers' Union)	J. Linahan (Labor)	J. M'Donald (Nationalist)
3	25	7

1921

J. Linahan (Labor)	J. M'Donald (Nationalist)
13	10

1924

J. M'Donald (Nationalist)	W.H. Nicol (Labor)
8	15

(source: *Colac Herald* 25 Oct 1920, 31 Aug 1921, 27 June 1924).

In a settlement of 60 or more eligible voters, the total number of votes cast (35, 23, 23) suggests a fairly weak level of political engagement, especially when a booth was opened at the mill specifically for the electoral convenience of its residents. This may have been due in part, however, to the difficulty of fulfilling the residential requirements to qualify to vote. A small fee was demanded of non-ratepayers wanting to enrol, and the time and effort involved in registration deterred many in an industry of high worker mobility (Bongiorno 1996: 82).

In this period the Australian Labor Party supported policies of employment security, the regulation of wages and working conditions, old-age pensions, and the White

Australia Policy. In 1916 the party had split over the issue of conscription, with prime minister W. M. (Billy) Hughes leading his supporters into a new Nationalist Party government (precursor to the United Australia Party, 1931, and the Liberal Party, 1945). At the federal level, this government struggled to reconcile the competing demands of tariff protection for manufacturers, and open markets for pastoral and financial interests.

Support for Labor candidates among workers at Henry's Mill was quite modest. The Nationalist candidate and sitting member, James M'Donald, gained between 20% and 43% of the vote, indicating a significant disaffection with Labor policies and a willingness to return M'Donald in 1924 for his fifth term. Although such a pattern may reflect a residue of support for conscription and broader economic concerns, state elections were more generally contested on local issues. M'Donald, for example, campaigned vaguely, but successfully, on a platform of supporting measures which were 'for the good of the community' (CH 22 Oct. 1920). The handful of votes in 1920 for the Farmers' Union candidate , J. Black, may have derived from seasonal mill workers who identified more strongly with the struggles of farmers and selectors than with Labor politics. It remains unclear, however, whether the different political preferences expressed through the secrecy of the ballot box, and the support for social and economic issues they implied, were prominent in the daily, face-to-face interactions of mill workers.

Industrial relations

The Victorian branch of the Amalgamated Timber Workers' Union of Australia (TWU) represented workers in both town and country sawmills. While some of the town mills were large businesses employing hundreds of men, most of the bush mills were small, isolated and dispersed. Union officials thus had the problem of representing fairly the interests of a diverse range of workers, as well as retaining members in numerous far-flung locations, many of whom moved in and out of the industry according to seasonal conditions (Day 1999: 161). This split constantly undermined the strength and authority of the union to negotiate on behalf of its members. TWU membership in country sawmills appears to have varied considerably. At some mills most of the workers were members, whereas very few were at others. This largely depended on the energy and effectiveness of the local union representative. In an industry blighted by injuries to workers, one of the main attractions provided by membership was disability and death insurance. In spite of the best efforts of organisers to cajole lapsed members and urge non-members to join, the union failed to gain state or national prominence until the protracted and unsuccessful Timber Strike of 1929.

Membership levels at the mills of W. R. Henry and Son varied considerably. At the No.2 Mill, only seven out of 15 employees were members in 1918 (TW 23 Sept. 1918). Specific figures for the No.1 Mill are not available, but monthly returns for the years between 1916 and 1923 suggest that only a dozen or two of the 60 workers at the mill maintained their membership. It is likely that many were committed to the union and paid their dues promptly, while others needed reminding and prodding, and some avoided it altogether. Three months passed quickly, and it was easy to forget or delay paying the five shilling fee when union officials were so rarely seen. Despite not having full membership, the No.1 Mill was regarded as 'white' by the union, as one where TWU conditions applied (TW 13 June 1916, 12 Oct. 1917).

Nevertheless, disputes arose between W. R. Henry and Son and the TWU. The union briefly forced the closure of the mill store in 1920, alleging that residents were forced to buy goods at the store, a claim denied by the Henrys (TW 31 Aug. 1925). In 1922 the firm was prosecuted on 30 counts of charging £1 per month rent for family houses, rather than the lower legal maximum of 12s 6d. In charging their married employees five shillings a week, split notionally in 2s 10d for rent and 2s 2d for grocery handling, they abided by the letter, if not the spirit of the law. The Henrys persisted with the higher charge in following years in defiance of union attempts to prevent them. They were also prosecuted for not paying their employees' wages fortnightly, as stipulated by union rules, but rather as the men wanted them. In this case the Henrys were forced to pay fines for infringing the law and suiting the convenience of their men (TW 31 Aug. 1925).

Timber workers were commonly employed on a contract basis. While labour in the mills was paid for in time wages, fallers in the bush were almost always paid piece rates. Most of the bullock teams hauling logs, and many of the horse teams taking timber to market, were owned by individuals operating as contractors to the sawmills. These were often local selectors supplementing their farm income with work at the mills (Dargavel 1995: 37). Paling and shingle splitters frequently contracted their loads to the millers in whose forest areas they worked. Only the more valuable employees were typically given the security of permanent employment. Engine driving and sawing, for example, were highly skilled tasks, and reliable men with these talents were hard to come by (McCarthy 1993: 38).

The TWU and the Commonwealth Arbitration Commission recognised more than 150 separate job descriptions in the timber industry during this period. Award conditions carefully stipulated the varying pay levels for each and every position. They distinguished, for example, between engine drivers and winch operators, bullock drivers and horse teamsters, head sawyers and No.2 benchmen, timber stackers and log yardmen, and between tallymen and ordermen (e.g. TW 6 Aug. 1915, 26 April 1921). It is unclear, however, the extent to which these distinctions were rigidly maintained in practice. When the industry was healthy and men were needed to work the mills, their ability to specify working conditions in relation to industry awards was enhanced. When demand was slack, however, and work scarce, a willingness to perform any task to hand may well have been the difference between keeping a job and losing it. Although differences between individuals with regard to union membership, employment basis, wage levels and job description may not have been significant in terms of belonging and worker camaraderie, they nevertheless remained a means by which the individual might define and distinguish himself from some, and identify with others.

Sectarianism

The dispute that erupted late in 1914 at Henry's Mill over the schoolroom behaviour of young Gladys Butcher was inflamed by the emotional correspondence of her mother, Amelia (see chapter 6: People). Among Mrs Butcher's array of complaints against the schoolteachers and other community members was her accusation that the children of other families were favoured on religious grounds at the expense of her own. She claimed that Francis Shine 'is the same religion a R.C....it is the protestant children he is always punishing'. She reported jearing the same teacher threatening to 'tickle the rotton [sic] wowsers up meaning the protestants' (VPRS 640/2423/3601, 10 Dec. 1915). Combined with her intolerance of drinking, gambling and smoking, Mrs Butcher clearly felt that she and her family were under siege by those members of the mill community who did not share or respect her family's religious allegiance.

Amelia Butcher's correspondence thus opens a small, local window on the bitter sectarian conflict that divided Australian society for so many years. Her attacks on the quality of the teaching and accusation of prejudice against her Protestant family also hint at the social and political aspects of the sectarian divide. By the early twentieth century, Protestants felt their values and world view to be at risk from a large and assertive Catholic minority. The historical working-class status and Irish origins of Catholics in Australia channelled them towards Labor politics. The broadly middle-class sympathies of Protestants, however, who supported individual enterprise and opposed socialism, meant they tended to support non-Labor parties (Broome 1980: 110-111; O'Farrell 1977: 277-81). The growth of Catholic churches and schools was also viewed with distrust by Australian Protestants.

The Catholic-Protestant divide extended to other areas of social life. Catholics, for example, regarded Sundays as a time for recreation, of picnics, sport and dancing, rejecting the common Protestant preference for a sober and decorous Sabbath (Broome 1980: 131-3). While the temperance movement was strong in both Catholic and Protestant churches, the former stressed the need for individual control, while the latter campaigned for the State to limit or prohibit the availability of alcohol (Campion 1982: 81-2). Gambling was widely regarded by Protestants as a social evil, but few Catholic consciences were troubled in the same way (O'Farrell 1977: 282-3).

Amelia Butcher's protests thus reveal the broader fears of Protestant sectarianism in this period. She clearly felt that her family, as non-drinking, non-gambling Protestants, was isolated among a community of drinking, gambling, Labor-leaning Catholics. She appears to have received little sympathy, however, fom her husband's employer and landlord. W. R. Henry not only opened his schoolroom to ministers of various denominations, but engaged workers with clear union and working-class allegiance. Although Anglican himself, Henry apparently had little in common with those employers who appended to their job advertisements 'No Catholics need apply' (Broome 1980: 98; Turner 1992: 55).

Ministers of religion would visit Henry's Mill once every two or three weeks, either Catholic, Baptist or Protestant. Ric Henry recalls that 'Everybody would go to the service, it didn't matter'. Significantly, however, he adds 'Of course, we were Church of England, so we didn't go to the Roman Catholic service, but we did to all the others. It was like that' (R. Henry 12 June 1999). Given the bitterness of the sectarian disputes in this period, it is unlikely that Catholics attended Protestant services, and vice versa. Attendance at the schoolroom on a Sunday morning may have served as a very visible expression of an individual or family's denominational allegiance.

Community

In spite of the potential for religious, political, social and industrial divisions, a sense of community could still find expression. Various picnics and sports days, for example, involving nearly everyone at Henry's Mill, were held over the years of the mill's operation. Teams representing W. R. Henry and Son played games of football, cricket and tug-of-war against Forrest and Barramunga, as did teams from Sanderson's Mill nearby (e.g. CH 7 April 1909, 26 May 1911). Henry's Mill also hosted an annual picnic for employees and former workers, usually held towards the end of January, when the skies had some chance of being clear. A large charabanc conveyed employees of the Geelong timber yard to Forrest, where they transferred to a timber tram for the journey to the No.1 Mill (e.g. CH 30 Jan. 1924). It was an opportunity for the town-based employees to meet, get to know and re-acquaint themselves with the bush mill workers and their families. The absence of clear and level ground precluded football and cricket matches, but permitted a range of log chopping events, including the standing block, underhand chop, and pairs and singles sawing (TW 16 March 1917). Donations of prizes from the Timber Workers' Union at such events encouraged wider participation among mill workers, whose working lives spent wielding axes and saws gave them a high level of proficiency on such occasions. The involvement of young boys was also encouraged, with a chopping event using smaller, lighter axes. While gratitude was expressed to the Henrys for providing the outing (e.g. CH 30 Jan. 1924), it was almost certainly the boarding house proprietress and other women of the mill who prepared dinner for the multitude.

A sense of belonging to the community was also expressed in physical ways. Exotic trees and shrubs, for example, were planted in the 'Port Arthur' residential area of the site. Surviving examples include two pines (*P. silvestris*), three oaks (*Quercus* sp.), a South American *Cestrum elegans*, and a *Hydrangea*. The larger of the remnant pines was probably associated with the school (Branditt 1922). Although generally fast growing, these specimens would not have advanced beyond a juvenile stage within the one to four year stay of most residents. They may have been planted by someone who lived at the mill site for longer than most, or by someone with the amenity of future households in mind. The exotics may simply have provided pleasure to those who planted and nurtured them, offering a familiar sense of 'garden' in an environment otherwise dominated by eucalypt regrowth, mud and weeds. Householders also cultivated garden plots to provide vegetables, while children grew plants and flowers in the school garden. In the face of a harsh environment, people did what they could to make the place feel like home.

The school building was also a potent physical symbol of the stability of the community, and the importance its members attached to educating their children. Squeezed in among the family houses, the building signalled the participation of the mill within the wider world, and the authority of the state within the lives of the residents. While the school had the immediate advantage of occupying the children and preventing them from 'running wild', it was also established as an incentive for men with families to settle at Henry's, thereby promoting, it was hoped, a more stable community and workforce. As well as having children gathered under its roof during the week, the schoolroom provided a focus for social events that helped bring mill residents closer together.

Charles F. Branditt taught at Otway Saw Mills between December 1919 and October 1922. Alone among teachers at the mill, Branditt appears to have felt some empathy for the community in which he worked. There is no evidence of him complaining to the Education Department concerning living or teaching conditions at Henry's Mill, and no requests for transfer. The reasons for his departure are not recorded. His portrayal of the mill in two articles contributed to the regional school magazine, *Forest, Lake, and Plain*, while romanticised, is nevertheless sympathetic to the trials and difficulties of work and life at the remote site (Branditt 1921, 1922). His positive response undoubtedly owed much to the presence of his wife, Nina, with him at the mill, and perhaps because, fresh from the Rural School Training College, he had no previous teaching experience to compare with his first posting. Although their accommodation arrangements are not known, Nina's annual income of £40 as the school sewing mistress meant the couple also enjoyed a higher household income than teachers before or after them.

Branditt records a number of social events that shed light on the participation of workers and residents in the life of the mill community. The School Committee, for example, formed a Saturday working bee to fall and transport a large tree to the school, where the older boys sawed and split it into firewood. Even though the men involved spent their working lives in such labour, and probably welcomed the weekend as a respite from it, their willingness to commit free time and labour reveals a sympathy for the children and the cold schoolroom they had to endure.

The rationale for subsequent community events at the school was a project to purchase books. This may well have been a common process in this era, when hundreds of small rural schools were established across the state to provide education for Victorian children. The Rubicon mill school in the Central Highlands, for example, was supported in 1925 by a donation of 200 books from the Children's Library League of Victoria (VPRS 640/P1/1671/4031, 23 May 1925). The Education Department was generally willing to meet requests for desks and blackboards, but unwilling to provide books for school children. This may have been due to financial stringency, or an approach to teaching that promoted rote learning over independent reading and thinking. Otway Saw Mills thus had to rely on social evenings on alternate Saturdays, and a school bazaar at Christmas, to raise money to acquire books for its pupils (Branditt 1922).

Saturday evening socials were held in the schoolroom, where the desks were removed, the floor was waxed, and music was provided by mill hands playing the accordion, harp and violin. Entry was one shilling. In the warmer weather, Nina Branditt brewed ginger beer, which was sold at a penny a glass, 'often returning a profit of ten shillings in one night'. The girls would also sell bunches of violets and other flowers they had grown in the school garden. Charles Branditt arranged guessing competitions involving silhouettes and mill residents and quizzes about Victorian country towns, for which prizes of d'oyleys, coat-hangers and tea-cosies were awarded.

Preparations for the Christmas bazaar were under way months before the event. The girls, taught by Nina Branditt, sewed prizes for the occasion. The boys raised young plants to sell to mill residents and made dolls' cradles from lolly boxes. They also erected an 'Aunt Sally', a painted figure of a woman smoking a pipe, and on the day offered small prizes for a knock-down at four shots for threepence. Mothers baked cakes and toffee, and made up jars of lollies and peanuts. The needlework pieces made by the sewing class were auctioned, before the schoolroom was cleared again for the dance. Almost £30 were raised to buy books.

Involvement in such school events had the potential to overcome other dividing loyalties of politics, class and religion within the mill community. The provision of school books may well have represented a cause in which all the residents felt they could make a contribution. It is unclear, however, to what extent the single men at the mill became involved. Lacking children of their own, some may have felt they had little to gain or contribute by participating in school-oriented events. Other, however, may have welcomed the opportunity for involvement, even vicariously, and to escape another evening in the cold, cramped huts they called home.

Charles and Nina Branditt also stand out for the leadership and energy they provided in promoting a sense of common purpose and unity at the mill settlement. None of the teachers before them or after them, however, ever made such an investment in communal life. In spite of their unique position at the mill, their access to the schoolroom and the moral authority they wielded, the reluctance or inability of most teachers to demonstrate leadership contributed little to any notion of identity, longevity or belonging at Henry's Mill.

Conclusion

The settlement at Henry's Mill, like most industry towns, was characterised by its ephemerality. It was never meant to last for long. Although the timber lasted for almost a quarter of a century, much longer than the eight-ten years originally proposed, the temporary nature of the mill was typical of the industry, and acknowledged from the beginning. It was created, like other temporary settlements, with a built-in obsolescence. The buildings and layout of the mill were of a short-term, expedient design, never intended to satisfy longer-term demands. It may, in fact, be more appropriate to regard the mill essentially as a place of work, where most people resided for shorter or longer periods, contributing somewhat marginally to a sense of communal belonging.

There is evidence, nevertheless, to suggest that a sense of community was fostered at Henry's No.1 Mill. The company held an annual picnic for its sawmill and timber yard employees, while teams representing the mill played football and cricket against the nearby townships of Forrest and Barramunga. The school and the education it provided to the mill children played a crucial stabilising role in settling families at the site, as well as offering a venue for dances, bazaars and religious services. Exotic trees and shrubs were planted for the pleasure of their appearance and the sense of attachment they imparted, where a house yard, at least, could be modified and claimed from the surrounding bush.

This sentiment of attachment should not, however, be overstated. The timber industry was renowned for the mobility of its workers, and Henry's was no exception. Individuals and households were continually arriving and leaving, always seeking better conditions, better pay, some improvement in their circumstances. Few people stayed at the No.1 Mill more than a few years, restricting any sense of common identification with the area. A range of tensions common to the broader community divided residents. One of the simplest social distinctions, marital status, received stark confirmation in the physical separation of single men's huts from married quarters. Some men were unionised, while others were not. A few enjoyed the security of permanent employment, but most worked on contract. Labor voters and Catholics were prominent, causing unease to the non-Labor Protestants. Generally, the tensions and disputes which inevitably arose in such an isolated little community remain hidden from view, only occasionally visible in terse court records and obscure newspaper columns. The people of Henry's, however, were by no means totally cut off from the wider world. They were as conscious as any of the often overlapping struggles of class, politics and religion, bringing these tensions and fears, as well as hopes and expectations, with them to their work and lives at the mill. This engagement with the world beyond the forest was also expressed in material ways. The commodities mill residents acquired and the meanings conveyed are examined in the following chapter.

Chapter 8

Consumption

Many material items were clearly considered necessary for everyday life in the Otway Ranges by the early twentieth century. Most of these objects were produced in distant industrial contexts rather than in the homes of consumers. At Henry's Mill, 19 different artefact fabrics were identified during analysis, representing at least 100 artefact forms, indicating the routine acquisition of industrial goods by forest mill residents. The limited income of timber workers nevertheless restricted their participation in mass consumer culture, challenging them to supplement store-bought goods with items acquired by a variety of other means.

This chapter traces the processes, practices and meanings of consumption at Henry's Mill. It begins with a brief overview of studies into the growth of mass production and consumption in the modern era, which has been variously cited in fifteenth-sixteenth, eighteenth and nineteenth-century Europe and North America. Several conceptual issues relating to consumer behaviour are then outlined. These include the separation of production processes from the context of consumption, and the various mechanisms by which consumers actually acquired commodities. It is noted that commodities, as objects of economic value, have meanings subject not only to their role in human transactions and encounters, but inscribed in their form and uses as well (Appadurai 1986: 5). Individuals assign meaning and symbolic value to goods within a framework of local understanding. In circumstances of economic hardship, these meanings and values can depart significantly from those intended by manufacturers, marketers and retailers.

The origins of a range of manufactured artefacts recovered from Henry's Mill are then examined, in order to assess the varying scales of economic interaction in which residents at the site participated. The processes of acquisition are also reviewed, including store purchase, direct delivery and itinerant hawkers. While the integration of residents within such broader exchange spheres was significant, their isolation required strategies to supplement building, dietary and other domestic needs, thereby maintaining rural traditions of resourcefulness and improvisation. Local production of various foodstuffs was thus important at the site, with vegetables cultivated, fish caught and rabbits trapped and hunted. Bricks and machinery components were reused in domestic contexts, while bottles were recycled into preserving jars. Assessment of the wages and general expenditures of mill residents is followed by a more detailed consideration of foodways, alcohol consumption, tobacco, tablewares and clothing. The material evidence supporting ideals of respectability is then reviewed, as a counterpoint to the ongoing pursuit of subsistence needs by the workers and residents of Henry's Mill.

Production and consumption

Mass production is intricately bound up with mass consumption. The manufacture and distribution of a vast array of new consumer goods in the modern period has decisively altered the material basis of human life. The origins of this 'consumer revolution', however, have been sought in different periods by numerous scholars (for critique see McCracken 1988: 1-30; Monod 1996: 102). Chandra Mukerji, for example, challenges the view that mass consumption was a product of industrial capitalism, identifying 'hedonistic' consumption in fifteenth and sixteenth-century Europe. She argues that markets were filled with silks, pottery, woollens and spices, while wealthy merchants and aristocrats adorned their houses with portraits, oriental rugs, tea services and furniture. The enormous social changes prompted by the growth of the capitalist economy included new domestic industries, fashions, and new cultural values for the use of new goods (Mukerji 1983: 1-9; see also Jardine 1996).

The consumer changes identified by Mukerji, however, were probably restricted to a limited stratum of European society. McKendrick and others have sought the growth of a consumer revolution in eighteenth-century England, where rising real incomes brought new products within reach of more people than ever before (McKendrick et al. 1982). Items once inherited were now increasingly purchased, with the dictates of fashion in clothing and pottery, for example, replacing the dictates of need. Luxuries came to be regarded as mere decencies, and decencies as necessities. Carole Shammas, however, is wary of such consumer 'revolutions', tracing instead the gradual spread of domestic commodities in England and America between 1650 and 1750 (Shammas 1990). She argues that 'groceries' (tobacco, sugar and coffee) and 'semi-durables' (textiles, pottery, glass and paper) all became items of prominent expenditure, used to enhance the domestic environment. She nevertheless acknowledges that European demand for these products had an enormous impact on national politics and social structure in the early modern period.

Rosalind Williams locates the advent of mass consumption in the late nineteenth century, as the necessary concomitant to industrial production (Williams, R. H. 1982). Focusing on France, she isolates ubiquitous publicity, large volumes of standardized merchandise, widespread reliance on money and credit, and the ceaseless introduction of new products as defining characteristics of the era. Increases in discretionary income meant workers could buy greater quantities of both staple products and non-essentials, as the torrent of technological changes lowered the cost of existing consumer goods and created entirely new ones. The enormous gains in productivity made available both more goods and more money with which to buy them (Williams 1982: 10).

David Monod has identified similar processes at work in late nineteenth-century Canada. The emergence of department stores and mail-order catalogues, along with changes in packaging, advertising and retail practice, accompanied by the expansion of railways and population growth, combined to stimulate supply and demand and the emergence of a new market economy (Monod 1996: 102-48). Mass marketing also transformed the purchase and use of goods in turn-of-the-century America, promoting the growth of a national market (Schudson 1984: 161-77; Strasser 1989: 15-16). Changes in labelling, advertising, distribution and retailing, in conjunction with rising incomes and cheaper manufactured products, turned customers into consumers. New needs and habits were created for which new products were marketed. This process involved a shift in the dominant value system, from one based on work, thrift and self-denial to one of leisure, spending and self-fulfilment (Jackson Lears 1983). Schlereth traces this explosion of consumerism in Victorian America, examining how millions were caught up in 'the sheer excitement of expanded consumption' (Schlereth 1991: 302). In a similar vein, Thomas Richards traces the growth of a commodity culture in later nineteenth-century England, where the advertising industry increasingly blurred the distinctions between producers, distributors and consumers (Richards 1990: 15-16). The benefits of mass consumption, however, were not available to everyone. Even at the beginning of the twentieth century, up to one third of the working class in England could not regularly afford enough food to maintain health and provide sufficient energy for work (Floud 1997: 100; Roberts 1971: 91).

Mass production, distribution and consumption of commodities was thus well established by the end of the nineteenth century. The emergence of this international capitalist economy depended in part on the development of new technologies in communication (telegraph), transport (shipping, canals, railways) and manufacture (textiles, tablewares, machinery, etc.). Ever-increasing quantities of fibres, minerals, grain and timber, along with consumables such as fruit, meat, sugar, tobacco, coffee, tea and cocoa flowed into Europe. Its rapidly expanding industries provided the means of exchange, with steam engines, mining equipment, textile machinery, steel mills, railways, ships, armaments and chemicals carrying the industrial revolution around the world. The domination of non-Europeans by colonial Europe increased dramatically, while gold discoveries in California and Australia multiplied the means of payment for commodities and encouraged the expansion of credit. In spite of economic lulls in the 1840s and between the 1870s and 1890s, the momentum continued until the onset of World War I, with the development of electric light and power, telephones, refrigeration, steel and steel alloys, turbines, and the internal combustion engine (Hobsbawm 1979: 31-8; Wolf 1982: 310-53).

Consumer meanings

One important feature of the growth of consumption has been a separation of the processes of production from the context of consumption. As the manufacture, distribution and use of goods intensified during the nineteenth and early twentieth century, people increasingly saw themselves as consumers, distinct from producers, identifying less and less with the institutions that provided goods and services. First with textiles, then ceramics, nails and bottles, increasing numbers of personal possessions came to be designed and made in distant places by unknown people (Miller, D. 1995: 31; Purser 1999: 123). Changes in the quantity, variety, complexity, form and use of these consumer goods influenced the ways people did their work and maintained social relations with each other. The transition from local production of domestic items to industrial manufacture meant that the material culture of even remote rural households and communities came to reflect the acquisition of mass-produced goods. People exchanged direct involvement in and knowledge of the production process for control over acquisition and consumer choice (Appadurai 1986: 41-6; Purser 1992: 105-7; Schlereth 1989: 347).

Another issue is the vectors of supply, the means by which so many diverse items came to be acquired by individuals. Analysis of the historical rise of mass merchandising and consumption after 1850 typically concentrates on the central role of advertising and department stores. This focus, however, marginalises other channels through which goods were obtained. Consumers of most social strata in Australia, North America and Europe acquired goods in open markets and fairs, directly from producers, or from travelling hawkers or peddlers. Many poorer consumers participated in the second-hand trade in clothes, household goods and machinery. Other means of acquisition included foraging for wild foods, home production of crafts and vegetables, barter, theft, gift and inheritance (Glennie 1995: 187).

A third issue is the *meanings* of material goods within their context of consumption. At the simplest economic level, people seek goods for subsistence, as well as for psychological comfort and competitive display. A much wider variety of cultural values and meanings, however, can be expressed through the acquisition of goods. Douglas and Isherwood have pointed out that although the goods themselves are neutral, their uses are social. That is, all material possessions carry social meanings, and establish visible and stable categories of culture. These shared meanings, however, are subject to drift and change, and constant renegotiation (Douglas & Isherwood 1978: 56-64; see also Bourdieu 1984: 100). Consumer objects communicate a wide variety of cultural values connected to notions of identity and social ideology, from luxury to self-sufficiency, extravagance to prudence, and vulgarity to respectability (Glennie 1995: 179-80).

Consumers may attach private meanings to goods quite different from the intentions of manufacturers and advertisers. In a process Daniel Miller has described as 'recontextualization', goods are transferred from their point of origin and appropriated within a locally meaningful system of values (Miller, D. 1987: 17, 167-75; Cook et al. 1996: 53-5). Purser has perceived such changes in consumption in nineteenth-century Paradise Valley, Nevada, where the local manufacture of coffins, for example, utilised such component parts as sawn timber, hinges and bunting. The result, a locally controlled material culture made from externally produced industrial items, required residents to negotiate their relationship to the wider world through the medium of consumer goods (Purser 1992: 110-111). Stewart-Abernathy has identified a similar process at work in early-twentieth-century Arkansas. Artefacts recovered from the Moser farmstead, such as canning jars, patent medicines, bottles and tablewares, were appropriated both as useful domestic devices and as powerful symbols of family independence within a supportive rural community (Stewart-Abernathy 1992: 114-20). Consumers thus acquired goods and constructed meaning in a purposeful fashion, employing strategy, creativity and self-reflection.

It is also important to consider the role of consumer goods within a context of minimal material comfort and social standing. Although the economic significance of working-class consumption may have been slight, it remained crucial to the identity and social interaction of working people. Material sparsity may even have intensified the personal meanings of particular items compared with richer consumers (Glennie 1995: 177). Households, communities and social groups which operated 'in the interstices of capitalism' frequently practised economic strategies which were kin-based, labour-intensive, little-capitalized and opportunistic. With exchange depending on barter as much as cash, a variety of income-producing strategies ensured economic security. People acquired goods they could carry away, and items they could mend. The archaeological signatures of such consumption should be read more in terms of functional flexibility, localised maintenance and portability, rather than in technological sophistication promoted by national advertising (Leone 1999: 14).

Markets

Analysis of artefacts recovered from the No.1 Mill revealed the origins of at least 71 manufactured items (Table 25). These provide some indication of the trade networks with which residents at the site were connected. Internationally, the main source of consumer goods was the United Kingdom. Artefacts included ceramic tablewares from the Staffordshire potteries and ammunition from London and Birmingham, along with Scottish firebricks and Irish whiskey. Only three items were identified from the United States: a Fig Syrup bottle from San Francisco, as well as a Barry's Tricopherous bottle and a Mrs Winslow's Soothing Syrup bottle from New York. In addition, ceramic dolls from Germany, a cup from Austria, a Dutch schnapps bottle, a Chinese teapot and three small amber bottles manufactured in Japan indicate wider import origins represented at the site.

Within Australia, the national market provided olive oil produced in Adelaide and patent medicines from Sydney. Melbourne, as an important national centre of manufacture, produced a wide variety of building materials, bottles, foodstuffs, chemicals, beverages, machinery, clothing and leather goods (Davison 1978: 41-71; Linge 1979: 232-3). Joseph Bosisto's eucalyptus oil, for example, was bottled in Richmond, and Greathead's Mixture was produced in North Melbourne (Arnold 1985: 68-9). Dr D. B. Jenner's Cough Balsam was manufactured by the Crown Chemical Company of Melbourne (Arnold, 1995: 31), while Felix Kruse's Prize Medal Magnesia came from Felton, Grimwade, & Company, also of Melbourne (Arnold 1987: 69). Melbourne-based condiment makers included the Rosella Preserving Company, Taylor Brothers Co-operative and Longmore & Company. Other items at Henry's manufactured in Melbourne included bricks, beer, and a stove from the foundry of D. & W. Chandler in Brunswick Street, Fitzroy. Regional Victorian towns were also important centres for the production of beer and aerated waters, represented at the No.1 Mill by J. J. Trait (Geelong), Bollington Hop Beer (Geelong) and Jonathon Reeve (Ballarat).

Analysis of the manufacturing origins of consumer goods at Henry's thus indicates that Melbourne was an important focus of production, with additional goods deriving from Geelong, Ballarat, Adelaide and Sydney. In contrast to the contemporary farming community at Silcott, where the majority of identifiable goods were manufactured in the major industrial area of the north-eastern United States, a greater proportion of goods at the No.1 Mill still derived from Great Britain. This reflected the continuing close economic and social ties between England and Australia, the comparative lack of economic ties with the United States, and the less developed state of Australian manufacture.

Purchasing

Residents at Henry's Mill employed a variety of strategies to acquire consumer goods. The most common method was to have goods brought out from Forrest along the timber tramline. Orders were sent down in an envelope with the engine driver to one of the two stores in the township. The following day the locomotive returned with all the orders, including quantities of flour, meat, tea and sugar, along with bags of chaff for the horses. Peter Evans describes a similar process in the Rubicon Forest, where orders from mill residents were dispatched with the timber on Monday and Thursday, and the goods brought up to the mills on Tuesday and Friday morning (Evans 1994: 110).

The Henrys also maintained a small general store for their employees, providing residents with the option of obtaining supplies on site. The store was mostly patronised by single men at the mill, who often used it to purchase small items such as tobacco and matches. In 1925 only one family out of ten obtained commodities from the store, while the others preferred to order from Green's or Archibald's in Forrest (TW 31 Aug. 1925; Sands & McDougall 1925). Several employees obtained pump trolleys for their personal transport, enabling them to get their own supplies in town (Houghton 1975: 56). The use of outside stores may have had a political dimension as well, with residents avoiding another layer of control exercised by the Henrys at the No.1 Mill. An official from the Timber Workers' Union was thus misreading the situation when he reported

Item	Manufacturer	MNI	Location	Country
brick	AR & FB Co	1	South Yarra, Vic.	Australia
brick	Knights	1	Lal Lal, Vic.	Australia
brick	Hoffman	2	Brunswick, Vic.	Australia
brick	Gartcraig	1	Ayrshire	Scotland
medicine bottle	California Fig Syrup Co.	1	San Francisco	U.S.A.
medicine bottle	Robert Greathead	1	Melbourne	Australia
medicine bottle	W. H. Comstock	3	Sydney	Australia
medicine bottle	Barry's Tricopherous	1	New York	U.S.A.
medicine bottle	Joseph Bosisto	2	Melbourne	Australia
medicine bottle	A. J. White	1	Sydney	Australia
medicine bottle	Curtis & Perkins	1	New York	U.S.A.
medicine bottle	D. B. Jenner	2	Melbourne	Australia
medicine bottle	Chamberlain	1	Sydney	Australia
condiment bottle	Rosella Preserving Co.	8	Melbourne	Australia
condiment bottle	Holbrook & Co.	3	Birmingham	England
whiskey bottle	Irish Imperial Whiskey	2		Ireland
whiskey bottle	Mitchell & Co.	1	Belfast	Ireland
aerated water bottle	J. J. Trait	3	Geelong	Australia
aerated water bottle	Bollington Hop Beer Co.	1	Geelong	Australia
condiment bottle	Jonathon Reeves	1	Ballarat	Australia
aerated water bottle	Cawsey Menck	1	Melbourne	Australia
condiment bottle	Taylor Bros. Co-op.	1	Melbourne	Australia
condiment bottle	Longmore's & Co.	1	Melbourne	Australia
bottle	United Glass Bottle Manu.	1	London	England
beer bottle	Volum Brewing Co.	2	Geelong	Australia
beer bottle	Abbotsford Brewery	1	Melbourne	Australia
medicine bottle	Felix Kruse	1	Melbourne	Australia
condiment bottle	The [Stonyfell? Oliv]E Co.	1	Adelaide	Australia
cup	'W'	1		Austria
toothbrush	Hygienic	1		England
shotgun shell	Eley	9	London	England
shotgun shell	Kynoch	1	Birmingham	England
saucer	Triple Crown China	2		England
plate	Ridgways	1	Shelton, Staff.	England
plate	J. & G. Meakin	1	Hanley, Staff.	England
plate	Pountney & Co.	1	Bristol, Glouc.	England
Plate	W. H. Grindley & Co.	1	Tunstall, Staff.	England
bottle		3		Japan
teapot		1		China
Jar	Solomon Solution	1	Melbourne	Australia
stove	D. & W. Chandler	1	Melbourne	Australia
Doll	Armand Marseille	1	Kopplesdorf	Germany

Table 25: Manufacturing origins identified at Henry's Mill.

in 1918 that conditions at the No.1 Mill were 'bad on account of the owner being the storekeeper' (TW 23 Sept. 1918). Living conditions were clearly challenging and difficult but residents had several options when it came to purchasing basic subsistence goods.

Forrest was the nearest practical retail centre for workers and residents at Henry's Mill. By the first years of the mill's operation, facilities in the town included two small stores, a butcher, baker, blacksmith and bootmaker, along with a school, post office, boarding house, hotel, Wesleyan Church and a local branch of the Commercial Bank (Sands & McDougall 1905). By 1910 the town also offered dining and tea rooms to serve the needs of travellers, and three new churches for the selectors and sawmill workers of the surrounding district (Sands & McDougall 1910). The general store of R. A. Lewis acted as local agency for the Royal Fire Insurance Company, and as produce and grain merchant for local farmers. The township existed to serve the immediate needs of the local community for accommodation, transport, basic commodities and commercial services. However, it was never to attract professionals such as solicitors or doctors, nor traders such as tailors, dressmakers or hairdressers. Colac or Geelong provided the nearest centre for such services.

With mail delivered six times a week to the No.1 Mill, residents may also have taken advantage of the catalogues provided by mail order houses to purchase goods directly from the suppliers. These 'universal providers' in Australia included Lassetter's (1902), Anthony Hordern & Sons (1897) and Foy & Gibson (1900). Manufacturers also published catalogues of their products, such as Harnwell & Son's Catalogue of stoves and ovens (*c.* 1920), and Edwin Jenkins, a Melbourne iron founder who advertised a range of household furniture (Jenkins *c.* 1890). It is generally very difficult to determine, however, the importance of catalogues in terms of purchasing, as many items available through mail order could be found in general stores as well. While mail order houses offered greater variety of goods at lower cost, orders were subject to delays, damage in transit, freight charges and often inferior quality.

Hawkers' vans were a common sight in rural Victoria during the later nineteenth and early twentieth century. Itinerant hawkers collected goods from a main distribution point and then worked their way out to towns and settlements along the track (Isaacs 1990: 114-17). Christine Stevens points to the Sikhs and Hindus who arrived in Australia late in the nineteenth century as the first of the 'Afghan' hawkers. She argues that they were probably from a merchant background and better suited to the individual pursuit of hawking than the more tribally-oriented Afghans who travelled the outback distances of South Australia, Queensland and the Northern Territory (Stevens 1989: 122). In the 1920s there were still bands of Indian hawkers trading to distant corners of Victoria (Bilimoria & Gangopadhyay 1988: 546).

Hawkers in the Otway Ranges were not necessarily always of non-European background. Fletcher Jones, early in the process of establishing a large and successful clothing business, bumped his way in a hawker's wagon between small Otway towns between 1918 and 1920 (Jones, F. 1976: 46-8). Sidney Myer was another successful retailer who began his career hawking in rural Victoria (Pratt 1978: 54-9). Sawmill settlements in Victoria may have provided an ideal environment in which travelling hawkers could operate. With pack horses or wagons they could penetrate to remote forest areas bearing small, relatively cheap and desirable consumables such as sweets, medicines, cloth and tobacco, permitting mill residents the indulgence of discretionary purchases to brighten up their otherwise drab living conditions (Kingston 1994: 22-3). It remains uncertain, however, whether hawkers ever made their way up to the No.1 Mill. Although 'Charlie Han' is remembered by Ric Henry as a 'cross between a Chinaman and an Indian hawker', he appears to have been an employee of rare, non-European background, one of whose roles was to divide meat cuts when they were delivered from Forrest (R. Henry 12 June 1999).

Foraging

The direct cash purchase of consumer goods was supplemented at Henry's Mill by various strategies of household production. These included vegetable growing and food preservation, fishing, hunting and the maintenance of a dairy herd. In the 1920s, for example, around 28 cows were kept for milking, grazed in the forest by Donald McPherson, a labourer at the No.1 Mill. Every cow had a bell around its neck, each with a slightly different note, permitting McPherson's collie dog to bring particular cows in for milking (R. Henry 12 June 1999). In addition blackberries were numerous in areas disturbed by logging and along watercourses, while honey from beehives was plentiful.

A stone alignment exposed during the excavation of House B apparently served as the border for a garden bed, while pines, oaks and shrubs represent the remnants of earlier plantings of exotics. With the slopes behind the huts and houses of the No.1 Mill mostly cleared of timber and heavy scrub, the rich mountain soil and reliable rainfall provided an ideal environment for growing vegetables. Gardens were a significant locus of food production which also communicated messages about household economy and personal responsibility (Rotman & Nassaney 1997: 58). Ric Henry recalls Oscar Henrikson, for example, maintaining a 'whopping big garden...[in which he] grew everything, everything you could eat' (R. Henry 12 June 1999). Most of the other people at Henry's Mill who lived in houses with a patch of land also grew their own vegetables, while the school children were encouraged to maintain garden plots in the school yard.

Fish commonly inhabited mountain streams in the vicinity of the No.1 Mill. Several of these made good eating, including the grayling, short-finned eel and blackfish, along with the introduced brown and rainbow trout (LCC 1976: 302-4). Many of the mill children were eager fishers, and made great sport out of spearing, catching or 'tickling' trout. They also ventured high up the Barwon River in search of minnows, scooping them up in nets made from flywire, before frying and eating them whole (R. Henry 12 June 1999).

Rabbits also provided an important supplement to the diets of No.1 Mill residents. The effective introduction of rabbits to Australia took place in 1859, not far from the mill. Thomas Austin released a mixed brace of wild and domestic rabbits at his property near Winchelsea, to provide hunting sport for English visitors. The rabbits quickly bred, and spread, throughout much of Australia (Barr and Cary 1992: 24-8). Although typically avoiding dense wet forests, rabbits nevertheless colonised the hillsides of the West Barwon valley disturbed by logging. They were hunted with guns, traps and ferrets, with excavation yielding the bones of at least three rabbits. Excavation of the three house sites yielded a total of eighteen .22 rim-fire cartridges, a single .410 shell and fourteen 12 gauge shotgun shells. Oscar Henrikson, for one, was a renowned gunshot who hunted rabbits and practised his shooting against the low-flying black cockatoos in the valley. A complete rabbit trap with a four-inch teethed jaw was also recovered from the surface of House A, while the mill children trained ferrets to catch rabbits as well. They placed nets over the entrance to burrows, put the ferret in and waited till the rabbits were chased out into the net (R. Henry 12 June 1999).

Reuse and recycling

Reuse and recycling of materials was evident both within the industrial context of milling operations and in the domestic confines of mill houses. Sawmills frequently acquired equipment, including saws, frames, boilers and engines, second-hand from mining companies, nearby sawmills or the Victorian Railways. This minimised the cost of purchase and often meant shorter distances to convey machinery to a new site. Once in place, however, considerable ingenuity was often required to adapt the milling plant and ensure its effective operation. The skills of a bush engineer were crucial in this process. With a basic toolkit of a hammer, chisels and spanner almost any task involving machinery could be undertaken (Houghton 1975: 42).

Alexander Sanderson established a mill in the Barwon Valley in 1900 which was powered by an ex-Victorian Railways locomotive, built 40 years earlier, to which he fitted flywheels and rope drives. The unit proved durable and continued in operation until the mill closed in 1923. Six other locomotives were also acquired by Sanderson to haul timber and logs, derived second, third or fourth-hand from a variety of sources (Houghton 1975: 40-3). W. R. Henry used two locomotives on his tramline. The first was originally built in 1889 for the Wallaroo and Moonta Mining & Smelting Co. in South Australia. Henry acquired it in 1911, but its lack of power prompted the purchase of a brand new machine from the Hunslet Engine Company of Leeds in 1912 or 1913 (Houghton 1995: 6-8). In later years the firm also built a motor-powered trolley for conveying mail between Forrest and the No.1 Mill. Its components included a four-wheel bogie fitted with a 600 cc 'Big Turn' motorcycle engine and a Harley Davidson gearbox and kick-starter (Houghton 1975: 56).

Bricks were another item commonly scavenged from other mills when they closed or relocated, or were purchased second-hand in small lots. At the New Federal Mill (1935-1949) in the Upper Yarra District, for example, bricks identified on the surface during archaeological survey included those made by the Hoffman, Montrose and Ordish brickworks. Nearby at the New Ada Mill (1931-1942), Hoffman, Gamble, Clifton, Fritsch & Holzen, South Yarra and Darby bricks have been recorded (Vines 1985: 149). Bricks appear to have been brought to Henry's Mill on a somewhat limited and occasional basis, illustrated by a request in 1913 for 100 bricks to repair the back of the school fireplace (VPRS 795/2812/3601, 27 Sept. 1913). Gartcraig firebricks were found in domestic hearths at Henry's, deriving originally from West Ayrshire in Scotland and presumably arriving in Australian ports as ballast in the holds of ships (Gemmell 1986:56). Knight's bricks came from Lal Lal, south-east of Ballarat, but must have been in circulation for some years before being brought to Henry's Mill, because the brick company went out of business in 1898 (Jack & Cremin 1994: 62). In addition, half bricks were retained and used in the construction of several features, including the open fireplace (context **37**) in House E, and the northern hearth (context **8**) in House A. This retention of whole and half bricks from a variety of sources reinforces their value and utility, in a context where available resources were curated and incorporated in the construction and maintenance of domestic features.

The northern hearth in House A also featured two iron rails, cut to length and fitted as a rough grate. Rails did not necessarily require sophisticated techniques to alter. They could be cleanly cut by making a deep nick around the piece with a hammer and chisel, before being dropped onto, or hit with, another metal object. Five short lengths of iron rail recorded in Houses A, B and E betrayed three different rail profiles, indicating their varied origins and the *ad hoc* nature of acquiring such materials. The incorporation of industrial items in domestic contexts is also indicated by an iron fire bar, salvaged from a boiler, inserted into the rear of the open fireplace (context **37**) in House E as a structural support.

Kerosene tins and boxes were also commonly reused and recycled at the mill. The recovery of glass lamp chimneys from House A, and fragments of at least seven kerosene tins, reinforces the importance of kerosene as a lighting fuel. When emptied, the four-gallon tins could be adapted in a wide variety of ways. Ric Henry recalls that when his mother wanted to bathe, several kerosene buckets full of water were heated on the open fire and then poured into an enamelled tin bath (R. Henry 12 June 1999). Kerosene tins could be made into wash troughs, cooking pans, watering cans, bread bins or flowerpots. Cut and flattened, the tin was valuable as a building material in walls and chimneys, while the wooden cases could be made into furniture (Isaacs 1990: 45-72). Empty tea and biscuit tins were also used for storing perishable food.

Sacks of flour, sugar, chaff and grains were regularly delivered to the No.1 Mill, where the empty fabric bags could be put to use as well. House walls, for example, were lined with cut-down hessian bags, over which sheets of newspaper were pasted to keep out the draughts (R. Henry 12 June 1999). Makeshift curtains and meat safes also employed hessian fabric. Beds were often made from two or three sacks with holes at both ends, where sapling poles were inserted on each side to form a frame. Calico bags, with a finer weave, were made into pillowslips, towels, peg bags, pot holders and horse blankets. In dire circumstances, the fabric was also made into clothing (McCalman 1984: 55).

Residents at Henry's Mill also altered and reused glass bottles. Eleven examples were recovered of beer, wine, soft drink and condiment bottles which had been cut down to form storage jars: seven from Dump C, three from House A, and one from House E (Table 26). Stuart has described two similar processes involved in reducing a beer bottle to a jar (Stuart 1993: 18). The bush method involved filling the bottle with water to the desired level. A string soaked in methylated spirit or other fuel was wrapped around the outside at the same level and set alight, causing the glass to crack in a horizontal line and creating a smooth break. Alternatively, a 'P' shaped iron ring was fitted over the neck of the bottle while the other end was heated. The heat was transmitted through the iron, causing the glass to crack.

The examples from Henry's have had the neck of the bottle removed at the shoulder, leaving an opening which varies from 55 mm to 70 mm in diameter, slightly smaller than the internal diameter of the bottle. The inside and outside edges of the fracture typically show tiny flake scars, which may result from the deliberate blunting of the sharp edge left by the neck removal (Stuart 1993: 19), or subsequent use and wear of the altered bottle. The lack of a formed rim on the resulting jar meant that a threaded, hinged or crown closure could not have been applied. Stuart notes that such items could not have contained a product under pressure, such as beer or carbonated drinks, as there was no means to hold the closure in place (Stuart 1993: 19). Instead, these jars were probably sealed with either wax or brown paper and flour paste.

A strong oral tradition suggests that the likely contents of such jars were jams, sauces or other preserves made from locally grown fruit and vegetables (Stuart 1993: 19). They could also have been used as vases for flowers. Unaltered bottles could also have been reused to store homemade sauces, ginger beer, etc. The garden plots in which vegetables were grown at Henry's expressed the skill, hard work and resourcefulness of the grower, while the preservation of seasonally abundant produce prevented wastage and supplemented food supplies. In a context of physical isolation and economic hardship, canning was also an important part of women's work and a source of pride and satisfaction. It provided the opportunity for display and competition in domestic skills, in terms of both the quality and quantity of food preserved. The recycling of bottles into jars is also consistent with other examples from the site of artefact reuse and adaptation, the archaeological signatures of self-reliance and improvisation. This kind of domestic production endured as an extremely important part of many Australian households well into the twentieth century, fitting into a culture and ethic of thrift and self-sufficiency.

Wages and expenses

Justice H. B. Higgins, president of the Commonwealth Arbitration Court, established a minimum basic wage in his Harvester Judgment of 1907. Shocked by the deep economic depression and chronic unemployment of the 1890s, Higgins sought to guarantee a minimum standard of living for all Australians. Designed to meet the needs of an unskilled working man, his wife and three children living in 'frugal comfort', he laid down a basic wage of seven shillings per day, or £2 2s per week. This was notionally to include ten shillings for housing, and 22s 3d for food, along with an allowance for clothing, transport, medical expenses, fuel and lighting (McCarthy, P. G. 1976). The judgment established a wage-fixing mechanism which was to be part of industrial arbitration in Australia for decades to come (Portus 1971).

Victorian timber workers in the early twentieth century earned wages generally a little above this minimum rate. Higher wages in the industry served partially to offset the many dangers and poor living conditions to which its workers were subject. At the beginning of World War I, the basic wage was fixed at about 50 shillings per week, while the timber industry award rate for bullock drivers and tree fallers was 60 shillings (TW 6 May 1914). By 1922, a mill hand at Henry's Mill earned around £5 per week, somewhat above the basic wage of £4 (BT 4 April 1922). In 1925, sawyers in the Otways earned around £6 10s, compared to the basic rate of £4 5s. A dispute at this

	House A	House B	Dump C	Sounding D	House E	Totals
Preserving jars	2	3			1	6
'bottle' jars	3		7		1	11
Totals	5	3	7		2	17

Table 26: Distribution of food storage jars.

time between a steward of the Timber Workers' Union and W. R. Henry & Son suggests that workers at the No.1 Mill were actually receiving wages above the award rates. One man left the mill when he realised his insistence on receiving the award rate meant his salary would actually be reduced (TW 31 Aug. 1925). Timber workers could also earn a little extra money for splitting tramway packing for the mill owner (Houghton 1992: 84).

Table 27 shows the weekly wages for different jobs in the timber industry in 1921. If the wage levels reflected the skill and value of different jobs, several observations can be made. Tree-falling, for example, was regarded as highly as driving a tramway locomotive or a team of bullocks. The highest paid position was that of a saw doctor, while a No.1 benchman earned more than such skilled artisans as millwrights and blacksmiths. Even 20-year-old workers, often with years of experience in the industry, were for the purposes of the industrial award, treated as 'boys' and paid accordingly.

At a time when suburban landlords in Melbourne charged 25 shillings a week for decaying three or four-roomed dwellings (McCalman 1984: 110), married employees at Henry's who paid five shillings per week for their small cottages enjoyed very cheap housing. The main expense for single men living in rent-free huts was board. In 1908 they paid 13 shillings per week for three prepared meals a day. By 1924 this had increased to 26 shillings per week, but salaries had also doubled (VPRS 795/1927/3601, 3 Oct. 1908; 640/2639/3601, 3 May 1924). Single men thus spent less than 30 per cent of their wages on food, which was purchased and prepared for them as well, leaving a considerable margin for tobacco, drink, clothing and other expenses.

Position	£	s	p
no.1 bench sawyer	5	2	-
no.2 bench sawyer	4	16	-
no.3 bench sawyer	4	10	-
docking bench sawyer	4	7	-
puller-out	4	4	-
saw doctor	5	10	-
locomotive engine driver	4	16	
winch driver	4	13	6
lever man	4	4	-
mill wright	4	19	-
blacksmith	4	18	-
tallyman	4	9	-
faller	4	15	-
mill labourer	3	18	-
bullock driver	4	15	-
boys <16 years old	2	6	-
16-17 years old	2	11	-
17-18 years old	2	16	-
18-19	3	3	-
19-20	3	11	-
thereafter	adult rates		

Table 27: Award wages for Victorian sawmill workers, 1921 (44 hour week; TW 26 April 1921).

Family men, with a wife and children to provide for, undoubtedly had much less discretionary income than their unmarried co-workers. Food alone absorbed a major fraction of their wages, even when modestly supplemented by home-grown vegetables and occasional fish, rabbits and blackberries. Families perhaps enjoyed, however, significantly higher levels of emotional support and psychological security. Teachers were arguably the worst off at the mill in financial terms, in spite of their higher education. Poorly paid, their salaries had to cover board, privately arranged lodging with a mill family, and in some cases, such as Cecil Wallis, support for a wife and child living elsewhere as well. The respect normally granted to those responsible for teaching the young may thus have been undermined by their lack of financial resources and tendency to leave the mill settlement at the earliest opportunity.

As tenants, mill workers had neither the opportunity nor incentive to invest in housing at the site. They were thus unlikely to expend any of their disposable income on the physical appearance of their dwellings. For a few this may have provided the impulse to save for future goals, itself a form of delayed consumption. Rotman and Nassaney suggest that [farm] tenants who placed little emphasis on modifying a landscape they did not own were more likely to embrace consumer culture to a degree which would belie their status as tenants (Rotman & Nassaney 1997: 56; see also Orser 1999: 153). Most married workers at Henry's, however, apparently enjoyed little in the way of a cash surplus to spend on goods beyond the immediate necessities and basic decencies of material life. Single men, with fewer responsibilities, may have enjoyed greater freedom in disposing of their income.

Foodways

Food consumption is influenced by many environmental, psychological, economic, cultural and historical factors. These include household income, individual preferences, availability, marketing, class and ethnic affiliation, as well as tradition, ideology and religion (Holt 1991: 50). A range of archaeological, historical and oral sources of evidence give some indication of the foodways practised at Henry's Mill. To feed a population of around 100 people, large quantities of food were delivered along the tramline from general stores in Forrest and Birregurra. Bulk commodities included sacks of flour, sugar, potatoes and onions, along with animal carcasses, large tins of jam and boxes of tea. The rapid improvement of canning technology in the later nineteenth century permitted a wider range of supplies to be ordered and stored as well (Busch 1981; Rock 1984). These supplies were supplemented at Henry's by various resources produced locally and on site, including dairy products, vegetables and rabbits.

Faunal evidence recovered archaeologically at Henry's indicates that cheap cuts of mutton were the most common form of meat consumed at the site. The basaltic plains to the north and west of the Otways were, and are, renowned sheep-carrying country. Supplies of mutton were readily available within the district. Rabbits were also hunted and trapped in the hill country around the mill, providing a valuable supplement to the meat diet. Beef cuts appear to have played a minor role in foodways at the site. Pigs were bred by the Simmons family, who farmed the western side of the Fork Paddocks on a curve of the West Barwon below the main tram tunnel. Perc Simmons cured the meat in his own stone smokehouse (R. Henry 12 June 1999). Although pig bone was rare in the recovered faunal assemblage, pork may well have been consumed as bacon, leaving little archaeological trace.

A herd of dairy cows grazed in the forest around the No.1 Mill by Donald Macpherson provided fresh milk and cream and the opportunity to churn butter. Chickens may also have been kept by individual households for eggs, but there is no evidence the birds were eaten. Vegetables grew well in the rich mountain soil, providing abundant produce on a seasonal basis. Numerous condiment bottles (n=28) indicate the frequent use of sauces, oils, extracts and essences. Most women at the mill baked their own bread in kitchen ovens, as well as scones and cakes. Occasional supplements to household diets included honey, blackberries and fish, gathered from the surrounding forest areas. The large numbers of ceramic cups and saucers recovered indicates the importance of hot drinks, especially tea, in the daily diet.

In a context of hard physical labour, geographical isolation and limited financial resources, food meant comfort, happiness and security. Historical photographs show, however, that very few timber workers were overweight. Although large meals were typical for workers in the industry, strenuous work and the need to maintain body temperature in a cold and often wet environment meant that calories burnt off quickly (see Conlin 1979: 167-9; Radforth 1987: 97). In spite of generic complaints by the Timber Workers' Union about the quality of food available in mill camps (e.g. TW 8 Dec. 1913), the evidence from Henry's indicates that workers and their families enjoyed a reasonably varied, ample and nutritious diet.

Alcohol

To the extent that the material recovered archaeologically from Henry's Mill represents a rough index of discarded domestic material, a total of 31 beer/wine bottles, three whiskey and two schnapps bottles from three households over almost a quarter of a century represents a relatively modest intake of alcohol. As there was no pub at the mill, and no evidence for on-site production of alcohol, little opportunity existed to refill bottles. Although patent medicines were infamous for their high alcohol content, and may have permitted a socially acceptable form of alcohol consumption, the presence of beer, wine and spirits bottles in the immediate vicinity of family houses and the mill school does not indicate a social sanction against drinking. There were more than enough legitimate health problems to deal with in an isolated mountain forest environment to justify the widespread use of patent medicines.

Conventional historical narratives of forest sawmills emphasise the drinking that went on in town pubs on weekends, while neither denying nor admitting the presence of alcohol at mill camps. Oral informants, however, are more forthright in acknowledging the role of alcohol at forest sawmills as a social problem. Archaeological evidence from Henry's Mill suggests that while various forms of alcohol were consumed in the family and single men's residential areas of the site, it was not drunk in large quantities. The following discussion reviews these divergent sources of evidence.

George Facey has produced several brief, chatty memoirs devoted to people, places and events he encountered in the Otways timber industry between World Wars I and II (e.g. Facey, G. A. 1977, 1981). The incidents he recounts concerning alcohol typically involve town hotels and/or binge drinking by workers on weekends. Norm Houghton has also pointed out how in the western Otways hotels were a 'perennial attraction on Saturdays', with pubs at Gellibrand, Beech Forest and Lavers Hill thronged with thirsty workers getting over a hard week's work (Houghton 1992: 85). Mill workers in West Gippsland visited town on weekends to 'let off steam after a week or more in the bush' (McCarthy, M. J. 1987: 57). Some workers in the forests of the Upper Yarra district went on drinking sprees, coming out of the bush every 3-4 weeks:

> They would first go to the barber in Powelltown for a bath, a shave, and a haircut, and change into their 'city clothes' which were kept in the barber's back room. They would then go down to the city, and after a week or so would return having spent their money. At the barber's they would change back into their bush clothes which they sometimes had washed and cleaned, and head back into the forest (Stamford, Stuckey & Maynard 1984: 74).

Information from informants and archives casts a slightly different light on the presence of alcohol in mill camps. While Joyce Harding, who lived at the Carisbrook Mill in the late 1930s, recalled that men 'mainly drank alcohol when they came out on weekends to Forrest' (J. Harding 21 May 1999), Jim Biddle, from the same site, remembers that there was 'always grog at the mills', and cites alcohol as the main reason for accidents in and around sawmills (J. Biddle 2 Nov. 1998). He recounts a story of W. R. Henry, who originally provided an 18-gallon keg at Christmas for the No.1 Mill workers, until the day when one drunk chased another around the mill with an axe, and the practise was stopped. Other tales circulated of the splitters and mill workers returning home on Sunday nights aboard tramway trolleys festooned with bottles and kegs (Houghton 1976: 7).

In terms of beer selection and consumption, only two different brewers were identified at the site: the Volum Brewing Company of Geelong, and the Abbotsford Brewery of Melbourne. Bottles from the Manufacturers' Bottle Company of Victoria, however, were widely used as an exchange bottle by many different breweries and were commonly recovered from the site. The absence of preserved paper labels undoubtedly obscures the range of brewers discernible. Nevertheless, the trend towards consolidation of production in the brewing industry during the late nineteenth and early twentieth century meant that less than a dozen different brewers operated in Victoria by 1920 (Deutsher 1999: 88-9; Lewis, M. 1992: 28). Contemporary advertisements in district newspapers also indicate the modest range of ales and lagers available to beer drinkers in this period. Consumers at Henry's thus probably had a restricted choice of beer produced by a limited range of brewers.

Richard Waterhouse has suggested that the working classes in Australia applied a high degree of discipline to their drinking, generally distinguishing between drinking and drunkenness, between sobriety and teetotaling (Waterhouse 1995: 82). Every stage of the timber-getting process was sufficiently dangerous that drunkenness or hangovers on the job endangered life and limb even further. Workers recognising the danger to themselves and their workmates may well have imposed their own sanctions about drinking on the job. At the inquests into the deaths of David Scott (1907) and John Brown (1926) the temperate habits of each was noted as a factor unlikely to have played a role in the deaths of either. The absence of a hotel at the settlement did not seriously inhibit the acquisition of alcohol by Henry's residents, but equally, the limited quantities recovered archaeologically also suggest that the option was not heavily or frequently taken up in the main residential part of the site. Bottled alcohol purchased at town hotels and intended for later consumption may also have commonly been consumed on the way back to the mill, with bottles discarded in the bush. Most alcohol consumption may, in fact, have occurred off site, at local town hotels, which bush and mill workers could visit at the end of the working week.

Tobacco

Among the 12 012 artefact fragments recovered from excavations at Henry's Mill, not a single clay pipe stem or bowl was identified. As clay pipes were inexpensive and fragile, and thus disposable, their archaeological recovery at the site could reasonably be expected if their use was common. A small circular tin recovered from House B, 62 mm in diameter and complete with a lid, is consistent with a container for loose-leaf tobacco, suggesting that the consumption of tobacco at the mill was overwhelmingly in the form of cigarettes.

After reaching its zenith in the 1890s, the popularity of pipe smoking declined rapidly in Australia, overtaken by the spread of cigarettes. Although more expensive than pipe tobacco, cigarettes were preferred because they were easier to light and tend, and because the smoke inhaled into the lungs had a greater physiological effect. For many men, rolling one's own became consistent with the conventions of masculine behaviour (Walker 1984: 54). Alex Larkins recalls that working men did not buy 'tailor made' cigarettes, but rolled their own from sealed tins of tobacco leaf, or carried a sharp knife to pare a tobacco plug (Larkins n.d.: 129). Clay pipes continued in production until the beginning of the First World War, after which they gradually disappeared (Matthews 1999: 169).

Tablewares

The identification of matching sets of ceramics in archaeological assemblages has important implications for determining the acquisition of tablewares within rural households and communities, and the social meanings of their use within such contexts. Tablewares could range from matched or closely matching sets to diverse, widely mixed collections. Households acquired tablewares according to a range of factors, including availability, cost, distance, fashion and need. The lack of published, accessible archaeological data from early-twentieth-century sites in Australia, however, means that comparisons with material from Henry's are difficult. Evidence from the United States, on the other hand, sheds some light on general patterns of tableware consumption.

Due to the fairly high rate of breakage in the transport of ceramics from Europe to America, the majority of patterns purchased by the ordinary consumer were common patterns and wares, representing items which could be easily replaced and were not susceptible to sudden changes in popularity (Majewski & O'Brien 1987: 179). Stewart-Abernathy, for example, has identified a high frequency of matched transfer-printed whitewares at the early-twentieth-century Moser farmstead in Arkansas (Stewart-Abernathy 1992: 117-18). He argues that sets were maintained because factory production of identical wares permitted replacement of individual pieces. The Moser families, in spite of their geographical isolation, were thus able to express a familiarity with contemporary fashions and participate in broader social rituals. Praetzellis and Praetzellis have documented the use by a later-nineteenth-century Sacramento family of white moulded wares in similar, but not identical patterns, to convey the impression of a matched dinner service. The collection, while relatively plain in appearance, was nevertheless fashionable, indicating a carefully conceived dining formality (Praetzellis & Praetzellis 1992: 91). Franzen's analysis of material culture from late-nineteenth and early-twentieth-century logging camps in northern Michigan identifies a similar predominance of white earthenwares among the ceramic remains. He suggests the use of such 'Ironstone' is not surprising since it was cheap, durable, and remained white even when chipped. The provision of ceramic tablewares, instead of enamel or tinwares, as well as fresh meat and cabins for families, were strategic ways for logging companies to attract workers during periods of labour scarcity (Franzen 1992: 86-94; see also Shackel 1993: 41-2).

Contemporary Australian mail order catalogues typically included numerous advertisements for fine ceramic tablewares. Often promoted under such names as 'Wattle' or 'Melbourne' to appeal to local tastes, the illustrated items generally featured transfer-printed or decal floral patterns around vessel rims and borders, with large areas left undecorated. Cheaper, undecorated ceramics, however, were usually included only in the small print of advertisements. Small country stores advertising in district newspapers tended to display few details of their available wares beyond 'Crockery at lowest rates'. In order to gain access to a reasonable selection of tablewares, residents of forest mills thus had to rely largely upon mail order catalogues or a rare trip into large regional centres.

The ceramic material from Henry's Mill was characterised by a large proportion of plain utilitarian white earthenwares. A total of 89 pieces (55.3% of total) were either plain or minimally decorated with simple moulded designs or gilt stripes. Transfer-printed or decal floral and foliate patterns were also common, ranging from simple flowers and leaves to complex polychrome roses. The remaining decorations consisted of an assortment of border bands and stripes, geometric designs, all-over coloured glazing and occasional scenic motifs. Documenting the prevalence of ceramic sets was difficult, however, due to the frequency of very similar white earthenware cups, plates and saucers, which were only distinguishable by slight differences in the placement of decorative motifs (e.g. gilt stripes) or in pattern designs. It remains uncertain how much variation each manufacturer, distributor and consumer tolerated in matching sets of ceramic vessels.

Of the 165 individual ceramic items identified archaeologically at Henry's Mill, 29 (17.6%) appear to have derived from exactly matched sets. These include seven cups, ten saucers and eight plates from House A, two saucers from House B, and two cups from House E. If minor variations in ceramic design and decoration are excluded, however, 92 close matches are identified. Many of these pieces thus complemented each other, giving the *impression* of a matched set. The most notable evidence for matched tablewares consisted of two dinner plates decorated with a black transfer-printed foliate pattern, with a matching tureen lid. The frequency of undecorated or minimally decorated white earthenware cups, plates and saucers suggests that residents commonly used matched or closely matched table and teawares which were cheap and sturdy. Such pieces would have been the least likely to show differences in patterns as there were no bright colours to stand out (Gaw 1975: 172). The remaining items, especially those with transfer-printed, painted or banded motifs, indicate that individual, more highly decorated ceramic pieces were also in use. Residents at Henry's Mill probably obtained ceramics from a variety of sources, including direct purchase or mail order, gifts and loans, inheritance, and the reuse of previous householders' goods, resulting in the use of closely matched tea and tablewares, as well as individual pieces. The prevalence of numerous closely matching table ceramics at the site indicates a general preference for simple, cheap and durable cups, plates and saucers.

Enamel or tinwares offered a clean, lightweight and durable alternative to table and kitchen ceramics during the later nineteenth and early twentieth century in Victoria. Only one enamelled vessel, however, was noted on the surface of Henry's Mill, in Dump C. It is uncertain whether such items were common at the site because their greater strength meant they were less likely to break and end up in discard deposits. Their use was consistent, however, with iron kitchen ovens (Gollan 1978: 96). Notably, enamelled cups, plates and other domestic items typically cost *more* in this period than the same items in ordinary white earthenware (e.g. Lassetter 1902: 508-12, 569). Residents at Henry's may thus have been making a simple choice of tablewares based on price. They may also, however, have preferred to serve food and drinks using heavier, more fragile ceramic vessels, emphasizing their membership in self-respecting society (Lawrence 2000: 132).

Clothing

Archaeological evidence also provides some indication of the clothing worn by workers and residents at the No.1 Mill. The 36 buttons recovered from Houses A, B and E appear to derive from various items of underwear, shirts, trousers, vests and coats. For the most part they were not gender-specific, and could have belonged to clothing worn by men, women or children. Contemporary photographs provide another source of evidence, indicating that felt hats or cloth caps were standard for mill workers, while leather aprons were commonly donned to protect clothing from rips, oil and grease (Houghton 1992: 88). Available evidence thus suggests that garments were sturdy, common and inexpensive, in a context where hard manual labour was the norm.

Several items devoted to the repair of clothing and footwear provide important clues regarding attitudes to physical appearance and the need to maintain personal possessions. A sewing needle, safety pin and scissor blade indicate the importance of mending garments, and perhaps the alteration of clothes belonging to older children to fit their younger siblings. Some clothes were probably also made from scratch. Fragments of boot leather, eyelets and a shoe last are reminders of the importance of *walking* as a means of local transport, the expense of footwear and the need to maintain it. Shoes and boots wore out only slowly in the bush because the ground was generally soft. The wet conditions and frequent mud, however, meant that stitching would sometimes rot out, requiring repair and the occasional replacement of heels and soles.

Residents also sometimes employed small items to enhance their outward appearance. A black facetted button from House A, for example, is suggestive of a more decorative style of dress. Press-moulded to resemble cut glass facets, it would have reflected the light, and probably derived from a 'fancy' men's waistcoat (Lindbergh 1999: 54). The wedge shank fastening made the button stand out from the garment, subtly enhancing its prominence. Two earrings and a mirror-glass fragment from House A, a brooch from House B and a cuff link and brooch from House E reinforced the idea that women and men consciously adopted contemporary notions of style and fashion into their modes of dress, and that these were acceptable and desirable expressions of femininity and masculinity (Lawrence 2000: 155).

The Archaeology of Respectability

A range of domestic items recovered from the excavated house sites at Henry's Mill sheds light on notions of respectability at bush camps, and the ways workers and residents negotiated the space between subsistence and decency. Pressed glass items from House A, for example, included a vase, two tumblers, an ice plate, and two sugar basins, hinting at concessions to comfort and modest refinement, their value perhaps heightened in a context of minimal household decoration. In addition, two painted ceramic figurines, a picture hook and furniture glass serve as reminders of the small touches which could brighten the drab, draughty interiors of dwellings. A tureen lid from House A indicates that certain foods, such as soups, stews or vegetables, were served at the table rather than ladled in the kitchen, suggesting the maintenance of dining propriety. Personal items and accessories had similar social meanings. The silver Albert chain from House B, for example, and the watch it once supported, was a symbol of the sober, punctual working man, having emerged by 1900 as 'a particularly male symbol of respectability and status' (Karskens 1999: 160). Women probably led the way, however, in maintaining standards of personal care and appearance. Two earrings, two brooches and a green perfume bottle attest to the concern of women at the mill for personal presentation.

Grace Karskens has suggested that domestic pride and effort were significant elements of respectability, a culture of behaviour transplanted to Australia from England by working and middle-class immigrants in the nineteenth century (Karskens 1999: 161). Clean clothing, personal cleanliness and domestic standards were vital to the achievement of a respectable life. An example of this concern at the No.1 Mill emerges in a letter from Mrs L. Riddle to the Education Department in 1925 (VPRS 640/2639/3601, 25 Feb. 1925). She complained that the Bush Nurse sent her children home from school because they had dirty necks and ears, when in fact the discolouring was only due to sunburn. She insisted that 'our school…is a model school in the bush…[and] every child is a credit to its parents'. Mrs Riddle was aggrieved that an outsider to the mill community so publicly (and mistakenly) questioned her domestic standards, humiliating her and her family, and jeopardising the respectability of the mill community more generally.

The experiences of Elsie Chamberlain also resonate with the pursuit of respectability, in circumstances where standards of cleanliness were difficult to maintain. Elsie and her husband, George, a mill hand, spent two years at Henry's Carisbrook Mill in the late 1930s, living in a rough wooden house. Elsie and her neighbour, Winnie Whitehead, were the only two resident women at the site. They paid 'proper' visits to each other, always knocking, and making sure their infant children were dressed in their best clothes. W. R. Henry usually visited the mill once a week and called on Elsie for afternoon tea. On these occasions she brought out her china cups, linen and special supper cloth. In all her time at the Carisbrook Elsie carefully followed her mother's advice, dressing each morning as if expecting company, even though almost nobody ever visited, and never allowing herself to become 'slummocky' (E. Chamberlain 7 June 1999). She also insisted on having a white picket fence built around the house yard, to reinforce her own views of domestic decency. Having grown up in Melbourne, however, it is uncertain whether the standards she maintained reflected those of mill residents more generally.

Archaeological and other sources of evidence thus point to attempts by some residents at Henry's Mill to create respectable family homes. Lawrence has explored a similar pattern among subsistence miners at Dolly's Creek in the later nineteenth century. Material items such as crockery and household furnishings reinforced the expression of respectable domestic environments. Living wages on 'a poor man's diggings' enabled the miners and their families to be not only self-sufficient but dignified and respectable on their own terms as well (Lawrence 2000: 127-43; 2003b: 219-21). Respectability was thus also related to economic status. Reliability of income eased the financial strain imposed upon family life by casual labour, and helped avoid financial destitution. Alex Larkins recalls that sawmilling and timber working during the early twentieth century in Victoria were respectable occupations, attracting good citizens who dressed a cut above labourers and other workers (Larkins n.d.: 20). Men who worked on contract, such as fallers or bullock teamsters, were typically fiercely independent and self-respecting. Mill sawyers and labourers earned steady incomes and the respect which came with hard work done well in a challenging environment.

Conclusion

Consumption at Henry's Mill was influenced by a range of meanings and values, including fashion, respectability, thrift, pride and hard work. The frequent acquisition of mass-produced goods, for example, indicates a routine engagement with industrial capitalism and an awareness of contemporary consumer trends. Although wage levels were modest, archaeological evidence suggests that they were sufficient to provide the necessities of daily life and occasional small luxuries. Workers and residents acquired commodities, however, in a variety of ways. These included patronising various stores, buying from hawkers, mail order catalogues, household production and scavenging. Several meanings can be inferred from these strategies. Residents avoided reliance on a single source of supply, and stretched limited incomes to ensure the best possible local value. As individuals and households already relied on W. R. Henry & Son for employment and housing, many may have been reluctant to submit to further authority from the firm, however symbolic, by utilising the mill store. The repair of clothes, reuse of bricks and the recycling of bottles, bags and kerosene tins at the mill also reinforces the thrifty, opportunistic nature of economic life at forest mill camps, as part of a wider rural culture of self-sufficiency and improvisation. The following chapter develops this theme further, analysing how the use of commercial medicinal preparations and other responses to injury and illness occurred within a context of self-diagnosis and healing.

Chapter 9

Health and Medicine

Residents of isolated forest sawmills in the early twentieth century adopted a range of responses to their health and medical needs. In the absence of readily accessible doctors, workers and families turned to proprietary medicines, homeopathic remedies and other personal cures. 'Patent' or proprietary remedies in particular were a popular form of self-medication during this period. Archaeological material recovered from the No.1 Mill included fragments of numerous patent medicine bottles, homeopathic phials and other non-prescription remedies. This material suggests that members of the community were part of a tradition of self-diagnosis and healing, one at odds with the emerging orthodoxy of scientific medicine and the rising authority of the medical practitioner.

Doctors in mid-nineteenth-century Victoria were poorly equipped to diagnose or cure their patients' illnesses. Treatment was based on a violent regime of bleeding and purging. Amputation was the most common form of surgery (Pensabene 1980: 6-32). By 1900, however, the germ theory of disease and its application to antiseptic surgery had revolutionised medical science, increasing the authority and standing of the medical profession. Advances in public health, including urban sanitation programs and the promotion of clean food and water, significantly reduced mortality rates. Doctors pursued their domination of the medical industry by marginalising other health providers such as midwives, herbalists and homeopaths, and tried to keep their patients firmly under 'doctor's orders' (Willis 1989: 68).

This chapter reviews the evidence of health and hygiene at Henry's Mill, emphasising the makeshift accommodation arrangements and variable quality of the water supply, and the implications these had for the origin and spread of illness. The industrial accidents and fatalities suffered by workers at the mill are also described, followed by a brief outline of the development of professional medical practice in Victoria in the later nineteenth and early twentieth century. The responses of mill residents to health problems are then assessed, including the high expense and limited availability of doctors in remote forest areas, the work of Bush Nurses, and the role of friendly societies. Material evidence for the use of patent medicines and other self-prescribed cures is also examined, demonstrating the need of residents to seek alternative treatments, their involvement in wider commercial networks, and the limited extent of medical influence on rural family life in this period.

Hygiene and health

Hygiene was an important issue for residents at Henry's Mill. Houses were built close together on the few available areas of relatively flat ground. They were generally structures of just a few small rooms. Walls were lined with hessian and newspaper to exclude draughts. Huts for single men lay in the path of a shallow, swampy gully draining into the West Barwon River. Each consisted of a single small room with a large open fireplace at one end for heating and cooking. Roofs were made of split palings or corrugated iron, and windows, where supplied, were small and fixed. The only source of cold food storage was a 'Coolgardie' safe, and lighting was provided by kerosene lamps and candles. A portable iron or enamelled tin bath provided bathing facilities for some, filled with water from kerosene tins heated in fireplaces. Toilets were simple earth pit privies, built close to the rear of houses.

Domestic water came from several sources. Some houses had galvanised iron roofs and rainwater tanks. The school also had a 600-gallon tank for the exclusive use of the pupils, but such a clean and accessibly supply was a temptation for their parents, who were accused by one teacher of stealing from it (VPRS 640/2328/3601, 12 Sept. 1913). A metal band recovered from below the tramline is consistent with a barrel hoop that may originally have been used to collect rainwater run-off from eaves. Water was also hauled from the river in kerosene tin buckets. When the mill first opened, the West Barwon River was described as 'a perpetual stream of the purest water' (VPRS 795/1927/3601, 9 Jan. 1905). However, logging operations higher up the valley, and illegal dumping of sawdust in the river, soon contaminated the flow. By 1921 the river water was generally unfit to drink.

Poor quality water increased the risk of gastrointestinal infection, while earth pit privies, built close to the rear of houses, raised the risk of typhoid outbreaks. Small, cramped houses made it difficult to isolate sick people, while the cold and damp climate aided the spread of chronic bronchial infections. Men in bush crews, who worked in the forest falling trees, clearing snig tracks and driving log trucks, often had to work all day in the rain in sodden clothes. These were difficult, if not impossible, to dry in the winter months. Colds and chest infections were presumably endemic to the working environment (Davies 2001b).

Teachers also complained about the wet weather and the effect it had on their health. William Morris requested a transfer in 1913 because the climate had brought on a sustained bout of influenza and neuralgia. Bernard Flood and Arthur Woods each requested transfers on the grounds of general ill-health brought on by the weather. The Education Department acknowledged that it was difficult to keep a teacher at the school during the winter, bemusing and frustrating the parents of school children, who had little choice but to endure the colder months and make the best of it.

The usual diseases of childhood were also present at the mill, such as chickenpox, measles and mumps. Head lice were also a problem (VPRS 640/2423/3601, 7 Dec. 1916). The highly contagious nature of these viruses meant that once one child contracted the disease, others would almost certainly follow. The school building was cramped, poorly ventilated and ill-lit, and provided an ideal environment for the spread of infection. Only one small vent in the ceiling admitted fresh air. Departmental policy required at least 12 square feet of floor space per child, but the 38 pupils attending by 1923 had to make do with 8.9 square feet each (Blake 1973 vol.1: 329; VPRS 795/2812/3601, 23 Feb. 1923). Although infected pupils were excluded from school for up to four weeks, the way the children lived in close proximity and played together out of school hours meant the exclusion was practically useless, and widespread infection was likely.

Early in the century hygiene played little or no part in school learning. Children were expected to learn clean habits by example at home. As part of the burgeoning public health movement in this era, however, school hygiene quickly became an integral part of the teaching curriculum in Victoria (Cumpston 1989: 120-5). The use of a common drinking cup and school towel ceased, recognised as potent sources of infection. The cleaning of slates by spitting on them and wiping with a sleeve was no longer allowed. Talks on health, correct posture, public behaviour and, for older children, the parts of the body, were recommended. Although schools were supposed to have been built according to contemporary standards of hygiene, the 'antiquated little school room' at Henry's Mill was clearly not up to the standards of the day.

More seriously, diphtheria remained a major killer of children well into the twentieth century. Although an effective antitoxin had been developed by von Behring in Germany in the 1890s, this was not always available in remote areas. As a result, fumigation persisted as a counter to the disease. Fumigation was a response to the miasmatic theory of illness, which held that fevers were related to 'toxic effluvia' rising from drains, cesspits and decomposing material. By the turn of the century scientists were learning that disease-causing bacteria do not travel far by air, leading them to question such measures as fumigation (Starr 1982: 190). Without ready access to effective treatment, however, these traditional precautions against 'miasmas' prevailed. At a remote sawmill in the Upper Yarra district in 1927, for example, a Mrs Leslie assured authorities that her efforts to ensure her children were free of diphtheria including fumigating the house 'all through with burning sulphur' (VPRS 640/P1/1671/4082, 16 June 1927). In the same area, a mill schoolteacher applied for a transfer in 1921 after one of his sons contracted diphtheria and another had died from it. The teacher's family had learned how tragically cut off they were from medical help, and 'the awfulness of isolation' (VPRS 640/P1/1671/4040/14 Aug. 1921).

Injuries

Timber workers risked injury at every stage of the timber-getting process. Falling trees and branches threatened bush crews, while a slip with a razor sharp axe could inflict serious wounds. Fingers inadvertently caught between logs and winch cables were quickly and painfully removed, while logs could roll unpredictably at landings or the log yard (Evans 1994: 121). Workers frequently suffered injured backs, and blood poisoning from splinters. Within the mill shed, safety equipment was rudimentary or non-existent. Neither safety helmets, gloves, goggles or earmuffs were supplied. The unguarded edges of high-speed saws sliced through fingers and limbs, while boiler explosions usually proved fatal to the engine driver and any others in the path of the flying debris. Exposed leather or rope belts driving mill machinery required constant care, while tramway accidents and derailments were also a frequent hazard.

The two fatalities known from Henry's Mill both involved tramline accidents. In 1907, David Scott, a post and paling splitter, was riding with another man, Frank Kelly, on timber trucks drawn by two horses. Going down a decline, the brake rope snapped. Kelly jumped clear, into a creek alongside the line. Scott, however, jumped the other way, landed against an embankment and fell back under the wheels. Terribly mutilated, he was pronounced dead from his injuries upon being removed to Forrest (CH 19 July 1907; VPRS 24/P/818/799).

In 1926, John Brown, employed as a horse teamster at the mill for the past 23 years, was crossing a bridge on a truck loaded with timber. The truck derailed from the bridge, and Brown was thrown into a gully below, suffering skull fractures and other wounds. He was transferred to the Geelong Hospital, but died from his injuries a few hours later, leaving a wife and five children (CH 15 Dec 1926; VPRS 24/P/1105/1346). Coroners' inquiries found that both Scott and Brown had died from injuries sustained through accident, and that no evidence of negligence or drunkenness was found.

Other injuries were reported at the mill over the years as well. Only months after David Scott was killed in 1907, Arthur Blair had his arm cut off (CH 27 Nov. 1907). In 1909, Arthur Battye's leg was badly crushed by a log, and Charles Ford lost a finger to a mill saw (CH 26 May 1909).

Leo O'Kelly, the schoolteacher, gashed his foot with an axe while chopping a limb from a tree. In 1913, an engine drawing four trucks was crossing a narrow bridge over a deep gully. One of the trucks, bearing W. R. Henry and his foreman, George Hall, was derailed. Hall was thrown over the bridge and fell 30 feet below, but crawled far enough away to avoid the truck which crashed down moments later. Henry suffered a badly injured foot, while Hall escaped with severe shock (CH 16 June 1913). Bushfires and snakes were also common threats to forest workers.

Medicine

Medical practice in mid-nineteenth century Victoria was widely characterised by rudimentary knowledge of disease, an inability to solve the health problems of the community, and a lack of public confidence in doctors. Diagnosis relied upon a superficial examination of symptoms, with treatment based on purging, bleeding and heavy doses of drugs. The lack of a secure, unified base of medical knowledge resulted in ongoing disputes between different schools of healing. Registered and unregistered practitioners of various types squabbled over qualifications and theories of treatment (Starr 1982: 81-102; Willis 1989: 42). Hospitals were the resort of the poorest and most desperate. Historian Grace Karskens has shown for The Rocks in Sydney that they were regarded as 'slaughterhouses' (Karskens 1999: 46). At the Melbourne Hospital, screams and cries from patients could be heard by anyone on the streets outside; 17 per cent of all cases admitted to the in-patient wards died at the hospital in the 1880s. Doctors competed with druggists, midwives, nurses, homeopaths and purveyors of various other treatments for supremacy over the medical treatment of the population (Finch 1998: 117; Pensabene 1980: 23).

The bacteriological discoveries of Louis Pasteur and Robert Koch in the 1860s and 1870s, however, lead to a much greater understanding of disease. Better control of water and food-borne diseases, and the advent of antiseptic surgery in the late nineteenth century, resulted in dramatic decreases in mortality rates. Advances in pathological anatomy and the germ theory of disease transformed medical practice and upgraded the public image of the doctor. Antiseptics and anaesthesia altered the community's perception of hospitals from death houses to life-saving institutions. Increasing levels of surgical intervention saw abdominal and orthopaedic surgery become widespread in Australian hospitals by 1900. By the 1920s almost every organ of the body was subject to surgery. For their part, doctors pressed their claims for professional autonomy, resulting in the increasing dominance of orthodox scientific medicine by the beginning of the twentieth century.

Demand for medical services also increased, especially among lower and middle-income earners who were increasingly willing to seek help for pain and illness. Fees for doctors nevertheless remained high, and alternatives were still sought by patients. Unregistered practitioners remained popular in Victoria well into the twentieth century, and included Chinese and Indian medical men, magnetists, herbalists, midwives and hydropathists (Martyr 2002). The greater accessibility and relative cheapness of chemists ensured their continuing role in dispensing proprietary remedies either over the counter or via mail order businesses. Travelling hawkers also sold medicinal preparations in remote areas. Homeopaths, osteopaths and chiropractors, while increasingly marginalised by 'scientific' medicine, retained an important presence in the Victorian health system until at least the 1930s.

Although many more medical conditions could be diagnosed accurately by this stage, the range of synthetic drugs available to treat diseases remained limited. A diphtheria antitoxin was developed in the mid 1890s, while salvarsan for syphilis was discovered in 1910 (Hooker 2000: 55-7; Shorter 1986: 92-7). Vaccines for typhoid and tetanus also date from the turn of the century, but cancer, tuberculosis, pneumonia, influenza and heart disease remained major killers (Report 1926). Many traditional diseases, such as smallpox, yellow fever, and to a lesser extent malaria, hookworm, cholera and plague, were gradually curtailed by 1920, due less to medical therapy than to the public health movement and its insistence on clean food, pure water and the removal of wastes (Shorter 1986: 119). Drugs to combat tuberculosis (consumption) and pneumonia were not developed until the 1930s and 1940s.

Measles, mumps and chicken pox remained obstacles for most children in this era, but no effective remedies were available. Mothers nursing children through such illnesses still frequently turned to patent medicines in search of help. These remedies were affordable, accessible and offered a reassuringly tangible response to medical problems. Most care of the sick remained in the home, with the burden of care falling upon the women of the household. Family medical manuals, such as Dr J. H. Kellogg's *The Home Handbook of Domestic Hygiene and Rational Medicine* and V. A. Wawn's *The Home Doctor* reinforced the tradition of self-treatment during the late nineteenth century. With such medical manuals, along with a few basic compounds available from the chemist, mothers would have to grapple with the everyday medical needs of their families.

Discussion

Doctors almost never made house calls to isolated sawmill camps. Ric Henry recalls 'There was no travelling doctors ever came up'. Nor was a telephone available to call for one. The nearest doctor for residents at Henry's Mill was at Birregurra, more than 40 kilometres away, or the hospital at Colac, more than 50 kilometres away. Timber trucks were used to transport mill workers needing emergency care. Those medics who did brave the bush elements in this period charged highly for their efforts. Hector Rule, a teacher at the Upper Britannia Creek mill in the Yarra valley, informed authorities that it cost ten to 20 guineas to get a doctor to within five miles of his school. Such expense made people hesitate, with possibly fatal results. The alternative, he argued, to transport a patient to a doctor through icy-cold and damp gullies in the depths of winter was 'a journey never to be forgotten' (VPRS 640/ P1/1671/4040, 14 Aug. 1921). When the young son of a tree faller in the west Otways was blown up by a detonator cap in 1925, a doctor was called from 40 kilometres away. By the time he arrived the boy had died. The doctor charged £12 12 s for his attendance, at a time when a faller earned less than £5 per week (TW 31 Aug. 1925). Poor quality roads in the Otways also hampered the transport of those who needed medical attention. When an employee of the Wye River sawmill was severely injured in 1920, it took seven hours to transport him by road to Forrest, a distance of scarcely 20 kilometres. He died several hours later (CH 3 Sept. 1920).

An alternative to a doctor was the Bush Nurse. These nurses attended patients, delivered lectures, and visited State Schools to examine students. Victoria's first Bush Nurse was appointed to the Otways township of Beech Forest in 1911. It was not until 1915, however, that a similar appointment was made at Forrest, offering practical nursing support to residents at Henry's Mill (CH 29 March 1915; Priestley 1986: 20-3). Children at the mill school were subject to a monthly visit from the Bush Nurse, who examined them for cleanliness and their general state of health.

In the absence of a doctor, residents at forest sawmills had to arrange for their own treatment. Pregnant women would leave the bush a month or two before their baby was due and board at a town nearby, returning two or three weeks after the birth. Women were also called upon to deal with horrific wounds resulting from logging and milling accidents. Ern Henry, son and partner of W. R. Henry, could also help with medical problems. Having trained as a chemist at the Ballarat School of Mines, he was able to concoct and dispense a range of medical preparations, and sew up cuts and gashes on horses and men. In the absence of antibiotics, however, the risk of infections from such injuries was acute. Wounds could only be swabbed with alcohol and dressed, in the hope that infection did not develop into septicaemia.

Another way of coping with injury and illness was membership of a friendly society. These societies originated in England in the seventeenth century, providing sickness and funeral benefits to members and a forum for social activity. In Victoria, the most prominent societies by the turn of the century were local branches of the Independent Order of Oddfellows and the Ancient Order of Foresters, along with the Australian Natives' Association. Societies also emerged on national (Irish) and denominational (Catholic, Protestant) lines. While the British societies only provided medical aid to the working man, the local Victorian societies offered medical coverage to wives and children as well. By the First World War, almost 30 per cent of the Victorian population had medical insurance with a friendly society (Green and Cromwell 1984; Ridley 1997). Disputes in the 1920s and 1930s between doctors and societies, however, saw membership plummet, as the cost of insurance increased, wage limits were imposed upon membership, the price of private medical care fell and the quality of treatment at public hospitals improved (Pensabene 1980: 149-57).

It is unlikely that many forest sawmill workers would have been members of friendly societies. Private patients typically received house calls from doctors, but lodge patients had to visit the doctor's surgery on designated nights, and received rushed, inferior treatment in the process. Doctors sought to treat a large number of patients in the minimum possible time, up to 30 or 40 an hour (Pensabene 1980: 151). Residents of isolated mill settlements were in no position to attend a doctor in the evening, and so would have received little medical benefit from society membership. Instead, sick or injured workers relied on insurance with the Timber Workers' Union, or an informal support mechanism among their workmates. Much of the attraction of union membership derived from the health and disability insurance it offered. As well, when a man was injured, the other men usually put in enough money to keep him going until he returned to work. Larger sums were raised when men were incapacitated. In the case of Arthur Blair, for example, £65 were collected after he had his arm cut off at the No.1 Mill (CH 27 Nov. 1907).

Homeopathy may have been another response to illness. Homeopathic practice enjoyed considerable popularity in Australia during the late nineteenth and early twentieth century. Developed by the German physician Samuel Hahnemann in the 1790s, homeopathy was based on the principle that 'like cures like'. Cures were effected by preparations which caused the symptoms of a disease to be duplicated, in weaker form, which the body could more easily overcome. The effects of drugs, it was believed, could be heightened by administering them in minute doses: the more dilute the dose, the greater the 'dynamic' effect. The effectiveness of homeopathy may have largely derived from allowing a disease to run its natural course, rather than interfering with the body's natural restorative processes. The very mildness of homeopathic therapies may also have contributed to their popularity among many women and their children during this period. Homeopathic preparations involved mixing powders or tinctures with water or alcohol, which were placed in small glass phials and shaken in special ways (Rogers 1998: 5-9).

It is difficult to trace the historical extent of homeopathic medicine because much treatment was self-administered. Homeopathic chests with instructions on usage were readily available, along with homeopathic guidebooks to 'self health', leaving much prescribing and treatment in the hands of the patient (Martyr 2002: 53-4). Archaeological evidence from Henry's Mill provides some indication that this form of medicine was being employed. Three glass phials were recovered from House B. One complete specimen, 73 mm in height and 14 ml in volume (i.e. half a fluid ounce, or four drachms), featured an intact cork closure. Six small, clear glass bottles were also recovered, circular in cross-section, 70 mm in height and one fluid ounce in volume (eight drachms). On the basis of their shape and size, and closely matching examples advertised in contemporary mail order catalogues, these small phials and bottles have been identified as containers for homeopathic medicines. These remedies were marketed as cures for almost as wide a range of ailments as patent medicines. Their presence at the mill suggests a willingness by residents to consider alternative treatments for illness beyond the strict orthodoxy of conventional medicine, and the continuing popularity of traditional approaches.

Residents also turned to patent, or proprietary medicines, to deal with illness. Such remedies had existed for centuries prior to 1800, but their numbers and use increased dramatically between 1850 and the 1920s, due in part to widespread advertising and frequent narcotic or alcohol content. Consumers were faced with a wide array of balsams, elixirs, essences, remedies and tonics, loudly and often fraudulently promoted as cures for up to several dozen ailments each. Although 'patent' medicine has become the common term for these medicines sold without a prescription, properly only those medicines with their ingredient formulae registered with the government were patented. More commonly, 'proprietary' medicines (or nostrums) were those drugs whose brand names or trademarks were protected by copyright. Lack of available or affordable doctors, and a persistent distrust of their treatments, ensured that patent remedies remained a popular alternative to orthodox medicine well into the twentieth century.

Some medicines were quite harmless, containing little more than sugar, colouring and flavouring. Others contained active ingredients which actually produced their advertised effect. Still others contained powerful drugs, including narcotics, in dangerous concentrations. Opiates were sometimes added not just to suppress pain but because heavy consumption could lead to addiction and thus increased sales (Parssinen 1983: 33). Chlorodyne, for example, was a cheaply obtained mixture of alcohol and opium, widely available throughout rural Australia, notionally intended for stomach and intestinal ailments. Proprietary teething powders were widely used to tranquillise crying feverish infants, and were typically based upon opium and chloride of mercury (Finch 1999: 78). Cigares De Joy (for asthma, bronchitis and hay fever) contained marijuana, while catarrh powders, a remedy for sore throats, contained up to four per cent cocaine. Hair restorers featured lead as an active principle, while Ayer's Sarsaparilla, probably the most widely advertised 'pick me up' of its kind in Australia, contained 26% alcohol (Report 1907: 92, 98). Mother Seigel's Syrup was also popular until the British Medical Association disclosed that it contained, among other ingredients, hydrochloric acid and treacle (BMA 1909: 176-7)

The first decade of the twentieth century saw patent remedies under growing attack at the hands of increasingly professionalised medical practitioners, who saw their livelihood under threat from providers of alternative remedies. In the United States, Edward Bok criticised the patent medicine industry in 1903 through the pages of the influential and widely read *Ladies' Home Journal* (Stage 1979: 140-62). Bok was quick to point out the hypocrisy of women active in the temperance movement habitually dosing themselves with alcoholic 'cures'. Samuel H. Adams exposed the deceptions of several hundred nostrum makers and quack doctors in a series of articles for *Collier's Weekly* in 1905. Legislators in the United States, Canada and Britain responded in the following years by gradually restricting the sale of opiate-based medicines (Carney 1981: 181; Starr 1982: 130).

In Australia, the crusading Octavius Charles Beale headed the 1907 Royal Commission on Secret Drugs, Cures, and Foods. His scathing attacks on the patent medicine industry played a significant part in the gradual reform of the drug industry in Australia. Beale's primary concern was women's use of drugs to prevent conception and induce abortions, thereby limiting family size and jeopardising the health of the nation and race. He attacked the secrecy and deception surrounding the ingredients of patent remedies and the ways they were marketed. He wanted to expose the dangers of these medicines by forcing manufacturers to print the contents on the label, encouraging informed use by consumers. Drug manufacturers resisted truth in labelling measures on the grounds that anyone would be able to reproduce their mixtures. They were supported by tabloid newspapers which derived large profits from advertising patent remedies (Report 1907: 428-31; Manderson 1993: 52).

Australian governments enacted various Poisons, Pure Foods and Opium Acts following Federation in 1901 to try to control the import, distribution and sale of opiates and other drugs. Victoria passed a Pure Foods Act in 1905 which required accurate labelling on the drug and alcohol content of medicines. Consumers, however, could still buy what they wanted as long as the drug was correctly labelled. In 1914 the Commonwealth prohibited the import of raw or prepared opium, and established strict licensing controls over companies dealing in medicinal morphine, heroin and cocaine. In spite of these regulations, there remained several dozen mixtures on sale which contained in excess of 0.2% of morphine. Victoria did not enact the provisions of the 1912 Hague Convention, dealing with government regulation of narcotics, until 1920 (Lonie 1979: 53-9).

The drug and alcohol content of proprietary medicines was thus subject to increasing regulation in Australia during the first decades of the twentieth century. Years of opposition from medical authorities resulted in tighter controls over

Name	Quantity	Advertised cure
California Fig Syrup	1	colds, headaches and fevers; constipation
Dr Morse's Indian Root Pills	3	'general antidote for all ailments'
Wood's Great Peppermint Cure	3	bronchial coughs and colds
Chamberlain's Cough Remedy	1	pneumonia and whooping cough
Joseph Bosisto's Eucalyptus Oil	3	rheumatism, sprains, coughs and colds
D B Jenner's Cough Balsam	2	coughs and colds
Greathead's Mixture	1	diphtheria, influenza, typhoid and other fevers
Barry's Tricopherous	1	hair and skin restorer
Mrs Winslow's Soothing Syrup	1	infant soother
Kruse's Magnesia	1	indigestion; laxative
Mother Seigel's Syrup	1	infant soother
Solomon's Solution	1	all purpose ointment
unidentified	19	

Table 28: Proprietary remedies recovered from Henry's Mill

the manufacture and sale of patent remedies. Nevertheless, medicines with significant narcotic and alcohol inclusions remained available. Australians remained avid consumers of patent remedies in this period, with the highest per capita consumption rate in the world. In 1905, such medicines accounted for almost one per cent of all imports into the Commonwealth (Finch 1999: 83).

Fragments of at least 38 proprietary medicine bottles were recovered from Henry's Mill. They were generally oval or rectangular in cross section, and ranged from 1.5 to 3.5 fluid ounces (50-100 ml) in volume. Although paper labels have not survived, 19 could be identified from inscriptions moulded into the glass. Nostrums found at the site were typically those marketed as providing relief from chest infections, and assisting with infants' and children's ailments. In spite of the high alcohol content of many patent medicines, they do not appear to have been consumed as a covert or socially acceptable form of alcohol at the site. There was no shortage of beer and wine bottles discarded under and around the mill houses, suggesting there was generally little need for discretion in the consumption of alcohol in the guise of medicinal preparations.

A bottle of Mrs Winslow's Soothing Syrup was recovered from House B at Henry's Mill. This syrup was the international bestseller of the proprietary soothers, and was widely used in Australia to calm teething infants and to cure diarrhoea. Until the end of the nineteenth century it contained about two per cent morphine, along with aniseed flavouring, alcohol and sugar (Finch 1999: 81). Opiate-based tranquillisers were popular because they stopped children crying, but their danger lay in children having a variable, and sometimes very low, level of toxicity to opium. Historian Lynette Finch has shown that such proprietary soothers were implicated in numerous infant deaths in Queensland in the 1880s and 1890s (Finch 1999: 79-85). Many other deaths were also alleged to be attributable to the misuse of such remedies (Carney 1981: 173). By the early twentieth century, however, morphine was removed from the mixture, so that it contained little more than sugar, water and four per cent alcohol.

Greathead's Mixture was a 'blood purifier', manufactured in North Melbourne and sold for 2s 6d a bottle. Its makers claimed it would cure diphtheria, typhoid, scarlet fever, influenza, quinsy, croup, bronchitis, coughs, colds and liver complaints (Arnold 1985: 68-9). California Fig Syrup, 'Nature's Pleasant Laxative', was promoted as 'ideal for children' and used to treat colds, headaches and fevers, as well as constipation (*Argus* 1 April 1899). Barry's Tricopherous was a remedy for 'removing dandruff, cleansing and restoring the hair, and for all ailments of the head' (Hechtlinger 1970: 37). Dr Jenner's Cough Balsam was aimed at all manner of chest afflictions, including whooping cough and pleurisy (*Argus* 2 July 1900). Felix Kruse's Prize Medal Magnesia was a popular remedy for stomach ailments, while a jar of Solomon's Solution was used to dress the wounds of 'cattle, horses, dogs and men' (Sands and McDougall 1904: 429). The diversity of patent remedies recovered from the mill suggests that residents were participating in the established western practice of self-medication. Their isolation from orthodox medical treatment meant they generally had little alternative to employing proprietary nostrums to aid digestion, soothe children, and ease coughs and colds.

Several other artefacts also reveal the medical responses of mill residents. A small glass prescription pill bottle, for example, was recovered from House A. While such bottles indicate a reliance on professional doctors and pharmacy, the single example from Henry's Mill appears to represent only occasional contact with orthodox medicine, in contrast to the dozens of patent medicine bottles and their emphasis on self-healing. There is also archaeological evidence for olive oil, eucalyptus oil and castor oil, along with fragments of kerosene tins. These may have been used in various bush cures. A popular remedy for winter colds, for example, involved gargling kerosene or eucalyptus oil. Alternatively, a few drops were mixed in a cup of hot water and sipped, or swallowed with sugar or olive oil. Numerous other cures for coughs were also tried, using such basic kitchen ingredients as onions, egg whites, glycerine, lemon juice and salt. An extract from manna gum (*E. viminalis*), a common tree species around the mill, could also be used as treatment for diarrhoea. Such methods and remedies were used at times in the home to treat illness without the involvement of orthodox medical doctors (Haggar 1979: 48; Thearle 1990: 126).

Conclusion

In her study on soldier settlers in Victoria between the wars, Marilyn Lake has pointed to the medical 'colonization' of the family, especially those living in remote areas in the 1920s and 1930s. She argues that increased surgical intervention and higher levels of medical surveillance in this period left the medical profession in a position of seemingly unassailable power, with families subject to 'doctor's orders' (Lake 1987: 151). Lynette Finch, however, citing coroner's records from late nineteenth-century Queensland, argues that in remote frontier settlements, parents themselves frequently sought medical service to cope with infant sickness. She suggests that parents in the nineteenth and early twentieth century typically reached into the home medicine cabinet for patent remedies, before worriedly seeking doctors to deal with serious infant ailments (Finch 1998: 126).

While orthodox medicine has left a more comprehensive written record, other, more private responses to illness and injury leave a strong material signature. The range of physical evidence from Henry's Mill indicates a variety of responses to health and medicine by residents of remote forest areas. The most direct response for most workers and families to the isolation, climate and crude living arrangements was simply to leave the mill in search of better conditions elsewhere. The timber industry in Victoria during this period was characterised by a high degree of worker mobility. For those prepared to stay longer, the cold and wet climate prevailing at the site meant that chest infections were common. Accommodation for families and single men alike was modest at best, with very limited space, poor sanitation and inadequate hygiene facilities. The great expense of doctors and their reluctance to journey out to the settlement meant that residents were effectively isolated from doctors' visits and hospitals except in the most extreme cases. In such circumstances they had little choice but to consider alternative treatments for medical ailments. Friends and neighbours were relied upon when injury or illness struck, and people dosed themselves and their children with proprietary medicines and homeopathic remedies. Such cures indicate a resistance to or lingering fear of orthodox scientific medicine, a preference for self-diagnosis and treatment, and a simple inability or reluctance to afford the expense of hospital or doctors' remedies.

Chapter 10

Conclusions

Some outside observers regarded forest mill camps as sordid and uninviting places. The Timber Workers' Union campaigned to improve the 'unkempt and unwholesome humpies' typically endured by its members (TW 10 June 1914). Justice Higgins, of the Commonwealth Arbitration Commission, lamented the 'slipshod' conditions of forest camps, with men 'pigging it' in huts and houses unfit for family life (Higgins 1920: 836). Were such judgments reasonable or accurate? Did sawmill workers and their families endure unusual hardships, or were their material and social lives comparable to those experienced by most other rural people of the era? This study set out to examine the character and conditions of forest sawmill settlements in early-twentieth-century Australia. Using Henry's Mill as a case study, different aspects of daily life at the site have been explored, to try to recall a way of life and work that no longer exists. A central concern has been to examine the responses of workers and residents to their geographical isolation, and the nature and extent of their interactions with each other and with the world beyond the forest. Within this framework, the study has assessed various elements of life at Henry's Mill, including living conditions, consumption patterns, and expressions of communal belonging.

Archaeological and other evidence from Henry's Mill indicates that living conditions at the site were indeed fairly basic. Huts and houses were small and dim, clustered close together and built from unpainted boards freshly sawn at the mill. Family dwellings were usually of three or four small rooms, with each house occupied by a succession of tenants over the years of the mill's operation. The single men's huts were tiny, one-room structures, usually shared by two workers. A pair of hearths revealed in each of the three excavated house sites indicates the importance of warmth and light in a generally cold and damp environment. In common with many other sawmills, the family houses at Henry's Mill were built several hundred metres from the huts of the single men, a spatial separation that reinforced the social distinction between the two groups.

The mill also featured a number of amenities for the convenience of workers and residents. W. R. Henry and Son operated a small, on-site store which generally provided small items such as sweets, tobacco and matches. Most families preferred, however, to order in their supplies in from Forrest along the tramline. The boarding house provided meals, but not accommodation, to mill workers. It played a vital role in alleviating the harsh conditions of bush life, with weekly costs deducted from the men's salaries. In addition, there was a post office, which handled mail deliveries six times a week. The school at Henry's Mill was also an important feature of the settlement. Not only was it the focus of education for the mill children, but it offered a venue for dances, fetes and weekly church services as well. Built in 1908, it opened in January 1909 and initially accommodated 16 pupils. By 1923, however, there were up to 38 children crammed into the ageing, 8 m by 4 m building. The school continued in operation until 1927, when the main mill shed was destroyed by fire.

The mill was built on sloping ground along the south bank of the West Barwon River, about ten kilometres south of the township of Forrest. The isolation of the mill meant that around 100 people lived permanently on site. Residents were not entirely cut off, however, from the outside world. The timber tramline provided a ready link to the railhead at Forrest and from there the world beyond. Nevertheless, gaining access to regional towns and cities was hampered by awkward railway schedules, and the need to allow several days for travel to such places.

W. R. Henry and Son employed a workforce of about 60 men to operate the No.1 Mill. There were several bush crews to fall trees and snig logs to forest landings, along with men to drive log trucks and others to work in and around the sawmill. Between ten and 20 of these men brought their families to live with them at the site. The other workers were either single men, or those who lived on site during the week and boarded their families elsewhere.

Children were also prominent at Henry's Mill. By the 1920s they formed up to one third of the mill's population. While school attendance occupied much of their day, there were also numerous chores to complete. Boys gathered and cut firewood, hauled water from the river and tended gardens. Girls cared for younger siblings, washed dishes and laundered and ironed clothes. When chores were done, a wide range of games and pastimes were enjoyed. Toys recovered archaeologically, such as marbles and doll fragments, indicate that children at the mill readily acquired mass-produced items, in spite of their physical isolation.

An important figure in the lives of the children was the school teacher. Extensive correspondence preserved in Education Department archives reveals a great deal about the responses of teachers to their work at the forest mill.

Fifteen different teachers taught at Otway Saw Mills during its 19 years of operation. All were male, and few were keen to stay any longer than necessary. Some complained about the cold and damp environment and its effect on their health. Others were unhappy about the accommodation available, and the cost of boarding a wife and child elsewhere. Several wanted to gain better teaching qualifications and so improve their employment prospects. Most applied for transfer on the grounds that the mill was just too isolated for comfort. Only Charles Branditt and his wife, Nina, seemed comfortable at the mill, and tried to improve the facilities available at the school.

Although teachers were poorly paid in this period, sawmill workers generally earned wages a margin above the basic wage, enough to meet the material needs of decent, everyday living. They purchased commodities that came from regional, national and international markets. While a few patent medicines derived from the United States, the United Kingdom was still the main source for such items as table ceramics, ammunition, whiskey and ink. Within Australia, Melbourne was the primary centre of manufactures, and produced a wide variety of building materials, bottles, chemicals, beverages and machinery. Regional Victorian towns were also important centres for the production of beer, aerated waters and condiments.

The remoteness of the mill site meant that workers and residents acquired goods in several ways. Grocery orders sent down with the timber trucks were filled by one of the two stores in Forrest and brought back out to the mill the next day. Alternatively, goods could be purchased through mail order catalogues, while travelling hawkers remained a feature of commercial networks in Victoria well into the twentieth century. Archaeological evidence indicates that a wide range of items were acquired, consumed and discarded at the site, suggesting a high degree of interaction with the world beyond the forest.

Nevertheless, thrift and self-sufficiency were still important rural values in this period. Many people at the mill cultivated vegetable gardens to supplement food supplies brought in from outside, while a herd of dairy cows were grazed in the forest to provide milk and cream. Children caught fish in the river and gathered honey and blackberries, while rabbits were trapped and shot in the hills around the mill. Cheap cuts of mutton were the most common form of meat consumed at the site.

Reuse and recycling were part of everyday life as well. Sawmills often acquired engines and machinery secondhand from mining companies or other mills. In the home, various items were scavenged and reused. Bricks and iron rails were built into fireplaces, while empty kerosene tins were fitted with a wire handle and made into buckets. Glass bottles were cut down and used as jars to store jams, sauces and other preserves. Children used sticks, pieces of rope, bottle tops and other items to create new playthings from everyday materials. It was these toys they had made themselves whose loss was most strongly felt when the mill shed was burnt out and the site eventually abandoned.

In spite of the hardships, life at the mill had its compensations. Rent was cheap and wages were steady. The Henrys appear to have been good bosses to work for. Food was ample, nutritious and reasonable varied. Alcohol was readily available, although apparently consumed at the mill in only moderate quantities. Glass tablewares, pictures and figurines were used to enhance drab domestic interiors. Personal presentation was important as well, finding archaeological expression in fancy buttons, watch chains and cufflinks for men, and brooches, earrings and perfume for women.

These general consumption patterns at the mill express the interaction of forest workers with a broader consumer society, as well as their physical isolation. Material evidence for medicines, ceramics, foodstuffs and numerous other goods indicates that the products of mass consumption were accessible and regularly acquired by mill residents. Commodities purchased, however, were generally cheap, portable and mendable. The repair, reuse and recycling of goods highlight the limited availability of certain items in remote areas, and the rural ethic of improvisation and thrift.

Material evidence also reveals private responses to medical problems, and the ways mill residents negotiated the tension between isolation and integration. Frequent wet weather, poor domestic hygiene and dangerous working conditions meant that injuries and illness were common. Water from the river was contaminated by effluent from the stables and sawdust from the mill, while earth pit privies were built close to the rear of houses. Men worked all day in the rain in sodden clothes, resulting in frequent upper respiratory and bronchial infections. Childhood infections such as measles, mumps and chickenpox were also present at the mill.

Workers and residents responded to these health problems in a variety of ways. Most care of the sick was in the home, usually in the hands of women, as doctors and hospitals were distant and expensive. Ern Henry, who had trained as a chemist in Ballarat, could also be called upon to dispense medicines and sew up wounds from axe injuries. A Bush Nurse paid monthly visits to the school children, inspecting them for cleanliness and their general state of health. Archaeological evidence reveals that homeopathic remedies and patent medicines were also a popular response. Most of the examples recovered from the site were originally sold as cures for chest infections and children's ailments. The popularity of such medicines at the mill reflects both the difficulty of getting a patient to a doctor, and the persistence of self-diagnosis and treatment in an age of increasing medical authority.

To what extent then, was Henry's Mill a 'community'? A community consists of a group of people living in close proximity and interacting together on a daily basis. Such people are united in some ways, divided in others, measuring individual and family goals against the wider needs of friends, neighbours and workmates. People at Henry's Mill were divided on the grounds of politics, religious denomination and terms of employment, but united by the camaraderie and satisfaction of hard work done well in a difficult and challenging environment.

Some sense of community was evident at Henry's Mill. Teams representing the mill played football and cricket against teams from nearby townships. The company held an annual picnic to foster ties between employees from its forest sawmills and the Geelong timber yard. The school provided a vital symbol of stability for families, and a venue for dances and religious services as well. People cultivated trees and gardens for the sense of attachment they imparted, and the feeling that a small space at least could be claimed from the monotony of the surrounding bush.

This sense of community, however, was often destabilised. Religious, social and political differences were commonplace. One of the main factors working against a sense of belonging was the mobility of the mill's employees. Few people stayed for more than three or four years, which meant most of the mill population was turning over and being replaced in that time. Married men with resident families stayed no longer than single men, undermining the idea that they were more stable and committed to their employment. People at Henry's Mill were always seeking better conditions and better pay, and were prepared to move to find them. Nevertheless, most remained long enough 'in the timber' for the mill to represent a transient kind of home in the bush.

Abandoned rural communities form a significant part of the twentieth-century archaeological landscape in Australia. Henry's Mill was only one of thousands of such places. Archaeologists, however, have been reluctant to engage with material culture of this period. This study has shown ways of entering a period at once familiar and strange, and how material evidence can help to construct knowledge of a past hitherto largely ignored. Archaeology can challenge conventional history by telling different stories about the past, and make visible those who were socially and geographically remote from mainstream affairs. Studies such as this must serve as a memorial to people and places already almost forgotten.

Appendix A

Excavation contexts

Context	Trench	Description
1	-	Surface leaf litter across site
2	A	Mixed light and dark brown soil, burnt clay
3	A	Dark burnt deposit in north-west of Trench A
4	A	Hard clay below 2 with orange and charcoal inclusions
5	A	Dark burnt deposit in centre of trench
6	B	Dark loose humus and leaf litter
7	B	Stone mound in south-east
8	A	Brick hearth feature north of 9
9	A	Brick hearth/oven foundation south of 8
10	B	Stone and brick rubble in grey clay
11	C	Pit: 1.20 m x 0.80 m
12	C	Pit: 1.50 m x 0.60 m
13	C	Pit: 0.65 m x 0.55 m
14	C	Pit: 3.20 m x 1.80 m
15	C	Pit: 0.70 m x 0.90 m
16	C	Pit: 0.70 m x 0.50 m
17	B	Loose dark brown humic soil in north-east of Trench B
18	B	Compact grey-brown clay
19	B	Compact yellow-brown clay and stone
20	B	Brick hearth structure in north-west of Trench B
21	A	Dark brown deposit below 3 and 4
22	A	Burnt wooden beams south of 9
23	A	Mixed dark brown deposit south-east of 9
24	B	Burning and metal fragments south of brick hearth 20
25	D	Gritty yellowish-brown deposit
26	B	Brown compact deposit below 24
27	A	Natural clay surface
28	D	Undisturbed gravel below 25
29	B	Loose dark humic soil within hearth 36
30	B	L-shaped stone alignment
31	A	Dark brown cultural deposit below clay of 9
32	B	Crumbly gravelly deposit south of 30
33	B	Crumbly dark brown deposit surrounding hearth 36
34	B	Loose rubble and soil on southern portion of 36
35	A	Northern portion of hearth feature 8
36	B	Brick hearth structure in south-east of Trench B
37	E	Brick and stone fireplace
38	E	Brick, iron and clay oven foundation
39	A	Make-up deposits of southern half of hearth feature 9
40	B	Crumbly, gravelly deposit in south-west of Trench B
41	B	Post-hole
42	B	Loose brown soil in ditch/depression west of hearth feature 20
43	B	Post-hole beneath 42
44	B	Test trench adjacent to hearth 36
45	E	Deposit above charcoal surface 46
46	E	Extensive charcoal deposit
47	B	Cobbled area below 17
48	B	Small cobbled area south of post-hole 43
49	E	Test pit in charcoal 46
50	E	Natural clay deposit below 49
51	B	West section of hearth 20
52	B	Brown gravelly deposit in east of Trench B above natural

Appendix B

The Residents of Henry's Mill

Surname	Firstname	Marital	Occupation	Arrival	Dept	Years
Allen	Joseph Edward	single	millwright	1917	1929	12
Bamford	Alice Christina	married	home duties	1919	1920	1
Bamford	Frederick George	married	labourer	1919	1920	1
Barton	William G		teacher	1926	1926	1
Beale	Eliza	married	home duties	1914	1918	4
Beale	William	married	labourer	1914	1918	4
Blackshaw	Donald		pupil	1905		0
Blackshaw	Ethel		pupil	1905		0
Blackshaw	Joyce		pupil	1905		0
Blackshaw	Helena	married	home duties	1905		0
Blackshaw	John Frederick	married	teamster	1905		0
Blair	Ivy May Windridge		pupil	1908	1908	1
Blair	Arthur	married		1908	1908	1
Blair	Tamar	married	home duties	1908	1908	1
Branditt	Charles Frederick	married	teacher	1919	1922	3
Branditt	Nina Berenice	married	home duties	1919	1922	3
Brew	Olive Millie Amelia		pupil	1908	1908	1
Brew	George	married		1908	1908	1
Brew	Mary	married		1908	1908	1
Brown	Lillian Margaret		pupil	1908	1910	2
Brown	John		pupil	1908	1910	2
Brown	May		pupil	1908	1910	2
Brown	Eileen		pupil	1908	1910	2
Brown	Ellen Winifred	married	home duties	1904	1926	22
Brown	John	married	teamster	1904	1926	22
Brown	John Thomas	single	labourer	1922	1925	3
Brown	Katie		pupil	1905		0
Butcher	Gladys		pupil	1914		0
Butcher	William Henry	married	mill hand	1914		0
Butcher	Amelia Jane	married	home duties	1914		0
Butcher	Clarice		pupil	1914		0
Butcher	William Henry Snellgrove		pupil	1914		0
Butcher	Mavis Clare Mary		pupil	1914		0
Carland	John	single	labourer	1919	1920	1
Castles	John	single	labourer	1913	1917	4
Cherry	Florrie May		pupil	1908		0
Cherry			infant	1908		0
Cherry	Amelia Jane	married	home duties	1908	1909	2
Cherry	William Henry	married	labourer	1908	1909	2
Close	Francis Albert	single	engine driver	1926	1927	1
Cole	William	married	mill hand	1919	1925	6
Cole	Florence Jane	married	home duties	1921	1923	2
Crosthwaite	John Lee	single	labourer	1922	1923	1
Culph	Frederick Arthur	married	blacksmith	1921	1928	7
Culph	Mary	married	home duties	1921	1926	5
Davison	George Henry	married	labourer	1925	1928	3
Davison	Nellie Jane	married	home duties	1925	1928	3
Day	George		pupil	1908	1910	2

Surname	Firstname	Marital	Occupation	Arrival	Dept	Years
Day	William		pupil	1908	1910	2
Day	Walter James		pupil	1908	1910	2
Day	Rupert		pupil	1908	1910	2
Day	Catherine		pupil	1908	1910	2
Day	Catherine	married	home duties	1908	1910	2
Day	John James	married	labourer	1908	1910	2
Dwyer	Alphonsus Leo		teacher	1924	1926	2
Edmends	Herbert	single	line repairer	1924	1926	2
English	W	unknown	school trustee			0
Evans	David Joseph	married	labourer	1925	1929	4
Evans	Gertrude	married	home duties	1925	1929	4
Flood	Bernard E		teacher	1913	1915	3
Ford			infant	1908		0
Ford	C	married		1908		0
Garland	Emily Louisa	married	home duties	1925	1929	4
Garland	William Francis	married	labourer	1925	1929	4
Garland	Stanley	unknown				0
Gilbert	Queenie		pupil	1908	1910	2
Gilbert			infant	1908	1910	2
Gilbert	Thomas John	married	labourer	1908	1910	2
Gilbert	Victoria	married	home duties	1908	1910	2
Gilmore	Emma Jane	married	home duties	1919	1925	6
Gilmore	Michael	married	labourer	1919	1925	6
Goudie	Herbert	married	labourer	1927	1929	2
Goudie	Lilian	married	home duties	1927	1929	2
Gray	Gertrude Janet	married	home duties	1921	1929	8
Gray	John	married	mill hand	1921	1929	8
Hall	George		pupil	1908	1910	2
Hall			pupil	1908	1910	2
Hall			infant	1908	1910	2
Hall	Elizabeth Ellen	married	home duties	1908	1910	2
Hall	Joseph George	married	bush foreman	1908	1913	5
Hall	Lillian May	married	home duties	1921	1922	1
Hall	Clifford Edwin	married	labourer	1921	1922	1
Hardy	A.	unknown	union steward			0
Harrison	Thomas Vernon		teacher	1917	1919	3
Hart	Horatio	single	labourer	1913	1914	1
Henderson	Alfred Edgar	married	mill hand	1924	1929	3
Henderson	Emmiline Eva	married	home duties	1924	1929	3
Henderson	Edwin Lawrence	single	mill hand	1928	1929	1
Henrikson	Eva	married	home duties	1926	1929	3
Henrikson	Oscar Albert	married	labourer	1926	1929	3
Henry	William Robert	married	timber merchant	1904	1928	24
Henry	Ernest Alfred	married	timber merchant	1907	1928	21
Hughes	Elizabeth Ann	married	home duties	1916	1918	2
Hughes	Richard	married	mill hand	1916	1918	2
Hughes	Edward James	single	mill hand	1921	1925	4
Icely	Amelia Mary Ann	married	home duties	1916	1917	1
Icely	John William	married	cook & baker	1916	1917	1
Irvine	William James	single	mill hand	1922	1925	3
Johnson	HROP	married	engine driver	1921		0
Johnson	Annie	married	home duties	1921		0
Keating	Vincent Denis	single	labourer	1925	1928	3
Kennedy	Herbert	single	faller	1915	1916	1

Surname	Firstname	Marital	Occupation	Arrival	Dept	Years
Kennedy	James	single	labourer	1916	1920	4
Killeen	Emily Lydia	married	home duties	1922	1928	6
Killeen	Thomas	married	labourer	1922	1928	6
Lindsay	Ellen	married	home duties	1912	1915	3
Lindsay	Harry	married	mill hand	1912	1915	3
Lindsay	W.R.	unknown	Union Steward			0
Lindsey	William		pupil	1908		0
Lindsey	George		pupil	1908		0
Lindsey	Leslie		pupil	1908		0
Lindsey	Reginald		pupil	1908		0
Lindsey			infant	1908		0
Mannix	Patrick	unknown				0
Matheson	Annie		pupil	1908		0
Matheson	Edith		pupil	1908		0
Matheson	John		pupil	1908		0
Matheson			infant	1908		0
McArdle	Caroline		pupil	1905		0
McArdle	Joseph		pupil	1905		0
McArdle	James		pupil	1905		0
McArdle	Caroline	married	home duties	1905		0
McArdle	Joseph	married	labourer	1905		0
McDonald	Donald William	single	labourer	1924	1925	1
McDonald	John	single	labourer	1924	1925	1
McGee	Leo P		teacher	1927	1927	1
McGlone	Rupert		pupil	1923		0
McGlone	Frank Lindsay	married	faller	1917	1920	3
McGlone	Herbert Edward	married	faller	1919	1920	1
McGlone	Olive Amelia	married	home duties	1921	1923	2
McGlone	Thomas John	married	mill hand	1918	1922	4
McGlone	James Henry	married	mill hand	1921	1922	1
McGlone	Claudia Mabel	married	home duties	1918		0
McGlone	Vidah Macrrena	single		1918		0
McGlone	T		pupil	1918		0
McGregor	Roy		pupil	1904		0
McGregor	May		pupil	1904		0
McLennon	L.	unknown	union steward			0
McMeikan	Jean	unknown	home duties	1912	1915	3
McPherson	Donald	single	labourer	1921	1929	8
Metcalfe	Florence May	married	home duties	1916	1917	1
Metcalfe	Robert Grange	married	mill hand	1916	1917	1
Miller	Frederick George	single	puller out	1916	1918	2
Miller	John Henry	married	labourer	1921	1925	4
Miller	Lily	married	home duties	1921	1925	4
Morris	William A		teacher	1912	1913	1
Moser	Leonard	married	engine driver	1916	1929	9
Moser	Jane Elizabeth	married	home duties	1922	1929	7
Myers	Edmund Robert	married	labourer	1913	1914	1
Myers	Frances	married	home duties	1913	1914	1
O'Connor	Edward C		teacher	1926	1926	1
O'Kelly	Leo C	unknown	teacher	1909	1909	1
O'Loughlin	Thomas Joseph	single	labourer	1925	1928	3
O'Neill	Charles		pupil	1905		0
O'Neill	Violet		pupil	1905		0
Parker	George Josiah	single	mill hand	1916	1918	2

Surname	Firstname	Marital	Occupation	Arrival	Dept	Years
Parker	Walter Stephen	single	splitter	1916	1918	2
Perkins	Herbert	single	labourer	1926	1928	2
Peterson	Charles Burbanks	married	labourer	1919	1922	3
Peterson	Minnie Eugania	married	home duties	1919	1922	3
Prendergast	Edward A		teacher	1915	1916	1
Reidy	Reginald Joseph	single	mill hand	1921	1925	4
Riddle	Robert Leonard	unknown	labourer	1916	1928	2
Riddle	L		parent			0
Roberts	Cyril	unknown			1921	0
Robinson	Eva	married	home duties	1916	1928	12
Robinson	Robert William	married	engine driver	1916	1928	12
Scales	Letitia		pupil	1908	1910	2
Scales	Frederick William		pupil	1908	1910	2
Scales	Violet Lillian		pupil	1908	1910	2
Scales	Esna Ann		infant	1908	1910	2
Scales	Edith May	married	home duties	1908	1910	2
Scales	Frederick William	married	labourer	1908	1910	2
Seebeck	Bernard Henry	married	head sawyer	1919	1928	9
Seebeck	Gustav Friederich	married	mill manager	1922	1929	7
Shine	Francis M		teacher	1916	1916	1
Smith	James Robert	single	labourer	1916	1920	4
Smith	J. G.	unknown	union steward			0
Stanford	Clifford James		teacher	1910	1912	3
Stanley	Richard	single	labourer	1921	1923	2
Stanley	May	married	home duties	1921	1925	4
Stanley	William Joseph	married	engine driver	1921	1925	4
Stevenson	Andrew Leslie		pupil	1905	1910	5
Stevenson	Andrew David	married	cook	1905	1910	5
Stevenson	Sophia Jane	married	home duties	1905	1910	5
Sykes	Alice Sarah	married	home duties	1921	1923	2
Sykes	Clarence William	married	blacksmith	1921	1923	2
Symons	Annie Elizabeth	married	home duties	1921	1925	4
Symons	Henry Charles	married	puller out	1921	1925	4
Templer	Eileen Vera	married	home duties	1925	1926	1
Templer	John William	married	labourer	1925	1927	2
Towers	Olive		pupil	1905		0
Towers	Millicent		pupil	1905		0
Towers	Amy		pupil	1905		0
Towers	Thompson	unknown	labourer	1905		0
Trickett	Ethel Laura Mildred	married	home duties	1924	1928	4
Trickett	William John	married	labourer	1924	1928	4
Turner	John Cecil George	married	engine driver	1908	1929	21
Turner	William Charles	single	labourer	1921	1925	4
Turner	Marjory W					0
Wallace	Sarah Ann	married	home duties	1926	1928	2
Wallace	William	married	mill hand	1926	1928	2
Wallis	Alfred Cecil	married	teacher	1922	1924	2
Walsh	Leonard B		teacher	1919	1919	1
Ward	Stanley Tunbridge	single	labourer	1926	1928	2
Wawn	R. C.	unknown	union steward			0
Wigley	Olive Annie	married	home duties	1924	1929	5
Wigley	Sydney Thomas Russell	married	labourer	1924	1929	5
Wilkinson	William Robert		pupil	1916		0
Wilkinson	Richard Joseph	unknown	ropeman	1916	1918	2

Surname	Firstname	Marital	Occupation	Arrival	Dept	Years
Wilkinson	George Henry	single	ropeman	1921	1925	4
Wilkinson	Henry John	married	engine driver	1921	1922	1
Williams	Edward	married	labourer	1921	1923	2
Williams	Ella	married	home duties	1921	1923	2
Wills	Arthur	single	labourer	1916	1918	2
Woods	Arthur W		teacher	1919	1919	1
Wright	Robert	single	labourer	1922	1925	3

Appendix C

(i): Artefact forms in House A by fragment quantity, minimum number counts and weight.

Form	Qty	MNI	Wgt (g)	Form	Qty	MNI	Wgt (g)
Architectural-fastening				*Industrial-fastening*			
corner fastener	3	3	4	bolt	2	2	885
hinge	4	4	690	nut	1	1	39
nail		1886	9646	spike	8	8	787
picture hook	1	1	1	split link	1	1	76
washer	8	8	52	*subtotal*	*12*	*12*	*1787*
subtotal		*1902*	*10 393*	*Industrial-hardware*			
Architectural-material				fish plate	2	2	1691
brick	2	2	7000	rod	4	4	635
masonite	12	1	13	*subtotal*	*6*	*6*	*2326*
window glass	59	2	119	*Industrial-hunting*			
subtotal	*73*	*5*	*7132*	rifle cartridge	2	2	1
Domestic-food				shotgun shell	12	12	39
bone	110	3	205	trap	1	1	757
mollusca	4	4	6	*subtotal*	*15*	*15*	*797*
subtotal	*114*	*7*	*211*	*Industrial-tool*			
Domestic-furnishing				chisel	1	1	18
battery	3	2	104	file	4	4	682
clock weight	2	1	8	wedge	1	1	355
figurine	11	2	26	*subtotal*	*6*	*6*	*1055*
handle	1	1	76	*Personal-accessory*			
hook	2	2	360	accessory	16	12	3.3
lamp glass	119	2	200	clasp	1	1	2
lock	3	1	10	earring	1	1	0.44
mirror	1	1	1	*subtotal*	*18*	*14*	*5.74*
stove	3	1	1489	*Personal-clothing*			
vase	5	1	121	button	20	15	13
subtotal	*150*	*14*	*2395*	eyelet	2	2	0.23
Domestic-kitchenware				needle	1	1	0.29
bowl	9	3	32	scissors	2	1	20
handle	3	2	126	shoe leather	7	1	21
jug	3	2	44	*subtotal*	*32*	*20*	*54.52*
subtotal	*15*	*7*	*202*	*Personal-hygiene*			
Domestic-medicinal				toothbrush	1	1	6
bottle	55	25	515	*subtotal*	*1*	*1*	*6*
glass slide	1	1	4	*Personal-toy*			
jar	2	2	200	doll	18	3	25
phial	6	5	63	marble	8	8	54
subtotal	*64*	*33*	*782*	model	1	1	20
Domestic-storage				toy tea set	2	2	11
bottle		54	11 866	*subtotal*	*29*	*14*	*110*
bottle stopper	10	10	86	*Social-currency*			

Form	Qty	MNI	Wgt (g)	Form	Qty	MNI	Wgt (g)
can	125	6	442	coin	2	2	11
cork	1	1	1	*subtotal*	*2*	*2*	*11*
crown seal	11	10	43	*Social-educational*			
jar	12	3	155	pencil	36	17	49
lid	5	5	51	slate board	7	2	34
screw cap	1	1	3	*subtotal*	*43*	*19*	*83*
tin key	2	2	9	Unidentified-fastening			
subtotal		*92*	*12 656*	banding	14	10	380
Domestic-tableware				subtotal	*14*	*10*	*380*
bowl	34	2	346	*Unidentified*			
cup	127	20	1531	bar	1	1	436
dredger	8	2	21	disc	5	5	108
lid	14	4	175	loop	2	2	25
plate	79	29	1309	plating	2	1	258
saucer	87	21	1285	ring	1	1	0.1
spoon	3	1	10	sheeting	20	3	297
teapot	4	2	61	spring	2	2	4
tumbler	7	2	51	tube	6	2	10
subtotal	*363*	*83*	*4789*	unidentified	1864		12 808
				subtotal	*1903*	*17*	*13 946.1*
				Total	7621	2263	59.121 kg

(II): Artefact forms in House B by fragment quantity, minimum number counts and weight.

Form	Qty	MNI	Wgt (g)	Form	Qty	MNI	Wgt (g)
Architectural-fastening				*Industrial-hardware*			
bracket	2	2	372	boiler pipe	2	2	1039
hinge	2	2	449	rod	2	2	162
nail		663	5267	*subtotal*	*4*	*4*	*1201*
washer	4	4	23	*Industrial-hunting*			
wood screw	2	2	8	rifle cartridge	16	16	10
subtotal		*673*	*6119*	shotgun shell	2	2	9
Architectural-material				*subtotal*	*18*	*18*	*19*
brick	3	3	9400	*Industrial-tool*			
flywire	28	1	1	file	6	6	1175
linoleum	4	1	2	punch/awl	3	3	371
mortar	5	5	12	shoe last	1	1	1176
window glass	13	1	17	*subtotal*	*10*	*10*	*2722*
subtotal	*53*	*11*	*9432*	*Personal-accessory*			
Domestic-food				bead	33	29	16
bone	42	4	117	brooch	3	1	1
subtotal	*42*	*4*	*117*	clasp	2	2	4
Domestic-furnishing				earring	1	1	0.6
handle	5	4	62	spectacle lens	1	1	2

Form	Qty	MNI	Wgt (g)	Form	Qty	MNI	Wgt (g)
hook	1	1	20	tin	1	1	28
lock	2	1	2.5	watch chain	1	1	4
stove	17	1	2203	*subtotal*	*42*	*36*	*55.6*
subtotal	*25*	*7*	*2287.5*	*Personal-clothing*			
Domestic-kitchenware				button	22	18	14
bowl	3	2	15	safety pin	1	1	1
handle	4	4	72	*subtotal*	*23*	*19*	*15*
jug	2	1	9	*Personal-toy*			
subtotal	*9*	*7*	*96*	doll	36	5	73
Domestic-medicinal				game piece	1	1	17
bottle	11	8	175	marble	18	18	120
phial	7	3	50	model	2	2	22
subtotal	*18*	*11*	*225*	pendant	1	1	3
Domestic-storage				toy tea set	7	4	27
bottle	782	17	5383	*subtotal*	*65*	*31*	*262*
bottle stopper	1	1	15	*Social-currency*			
can	7	1	93	coin	2	2	8
crown seal	2	2	9	*subtotal*	*2*	*2*	*8*
jar	9	3	149	*Social-educational*			
screw cap	3	1	5	pencil	14	11	31
subtotal	*804*	*25*	*5654*	slate board	4	1	33
Domestic-tableware				*subtotal*	*18*	*12*	*64*
cup	22	4	282	*Unidentified-fastening*			
egg cup	1	1	2	banding	7	2	105
plate	17	5	236	clamp	1	1	24
saucer	13	8	253	strapping	4	2	200
spoon	4	3	30	*subtotal*	*12*	*5*	*329*
teapot	1	1	4	*Unidentified*			
subtotal	*58*	*22*	*807*	disc	3	3	27
Industrial-fastening				loop	1	1	11
chain	4	4	1758	ring	2	2	13
nut	1	1	17	sheeting	32	4	126
spike	2	2	271	tube	8	2	20
split link	1	1	8	unidentified	344		4695
subtotal	*8*	*8*	*2054*	wire	1	1	1
				subtotal	*391*	*13*	*4893*
				Total	3149	918	36.360 kg

(iii): Artefact forms in House E by fragment quantity, minimum number counts and weight.

Form	Quantity	MNI	Wgt (g)
Architectural-fastening			
corner fastener	1	1	3
nail	361	230	1229
subtotal	*362*	*231*	*1232*

Form	Quantity	MNI	Wgt (g)
Architectural-material			
brick	2	1	27
window glass	9	1	27
subtotal	*11*	*2*	*54*
Domestic-food			
bone	115	3	108
subtotal	*115*	*3*	*108*
Domestic-furnishing			
handle	1	1	2
subtotal	*1*	*1*	*2*
Domestic-medicinal			
bottle	3	2	172
phial	1	1	14
subtotal	*4*	*3*	*186*
Domestic-storage			
bottle	358	15	5702
jar	15	1	123
subtotal	*373*	*16*	*5825*
Domestic-tableware			
bowl	26	1	104
cup	13	5	161
plate	8	3	74
saucer	8	5	60
subtotal	*55*	*14*	*399*
Industrial-fastening			
chain	1	1	1194
spike	2	1	44
subtotal	*3*	*2*	*1238*
Industrial-hunting			
shotgun shell	1	1	5
subtotal	*1*	*1*	*5*
Personal-accessory			
bead	2	1	0.62
brooch	1	1	2
cuff link	1	1	2
subtotal	*4*	*3*	*4.62*
Personal-clothing			
button	3	3	3
subtotal	*3*	*3*	*3*
Personal-toy			
marble	1	1	3
subtotal	*1*	*1*	*3*
Social-educational			
pencil	3	2	5
subtotal	*3*	*2*	*5*
Unidentified-fastening			
banding	14	1	84
strapping	8	1	38
subtotal	*22*	*2*	*122*

Form	Quantity	MNI	Wgt (g)
Unidentified			
disc	5	4	11
loop	2	1	13
plating	1	1	19
ring	1	1	1
unidentified	151		1361
wire	1	1	4
subtotal	*161*	*8*	*1409*
Total	1123	292	10.596 kg

(iv): Artefact forms in Dump C by fragment quantity, minimum number counts and weight.

Form	Quantity	MNI	Wgt (g)
Domestic-kitchenware			
bowl	2	1	20
subtotal	*2*	*1*	*20*
Domestic-medicinal			
bottle	4	3	232
subtotal	*4*	*3*	*232*
Domestic-storage			
barrel hoop	1	1	329
bottle	72	38	13 739
subtotal	*73*	*39*	*14 068*
Domestic-tableware			
cup	3	2	52
subtotal	*3*	*2*	*52*
Social-educational			
slate board	1	1	23
subtotal	*1*	*1*	*23*
Unidentified			
unidentified	1	1	5
subtotal	*1*	*1*	*5*
Total	84	47	14.400 kg

(v): Artefact forms in Sounding D by fragment quantity, minimum number counts and weight.

Form	Quantity	MNI	Wgt (g)
Domestic-storage			
bottle	15	6	2152
subtotal	*15*	*6*	*2152*
Total	15	6	2.152 kg

(VI): ARTEFACT FORMS FROM SURFACE BY FRAGMENT QUANTITY, MINIMUM NUMBER COUNTS AND WEIGHT.

Form	**Quantity**	**MNI**	**Wgt (g)**
Domestic-medicinal			
bottle	4	4	425
subtotal	*4*	*4*	*425*
Domestic-storage			
bottle	1	1	6
subtotal	*1*	*1*	*6*
Domestic-tableware			
plate	4	2	41
saucer	2	2	35
subtotal	*6*	*4*	*76*
Unidentified			
unidentified	9		26
subtotal	*9*		*26*
Total	20	9	533

Bibliography

Archival sources

Australian Archives

B5846/1: Victorian Post Offices

Box 39: 'Otway' file – Otway Saw Mills

Public Record Office of Victoria

VPRS 24/P: Inquest Deposition Files

Unit 818:799 – David Scott

Unit 1105:1346 – John Brown

VPRS 28/P3: Probate and Administration Files 1841-1960

Unit 4930: 415/726 – Probate Papers of William Robert Henry

VPRS 292: Birregurra Court of Petty Sessions Registers 1872-1955

Unit 5: 26.10.1898 - 23.1.1907

Unit 6: 23.1.1907 - 16.8.1916

Unit 7: 13.9.1916 – 9.5.1923

Unit 8: 6.6.1923 – 3.11.1927

VPRS 640: Primary Schools Correspondence Files 1878-1962

Unit 1671: Rubicon School No. 4031; Upper Britannia Creek School No. 4040; Mississippi Saw Mills School No. 4082.

Unit 2245 [Otway Saw Mills School No. 3601: 1907-1910]

Unit 2249 [Otway Saw Mills School No. 3601: 1907-1910 – supplementary]

Unit 2328 [Otway Saw Mills School No. 3601: 1911-1913]

Unit 2423 [Otway Saw Mills School No. 3601: 1914-1916]

Unit 2498 [Otway Saw Mills School No. 3601: 1917-1920]

Unit 2563 [Otway Saw Mills School No. 3601: 1921-1923]

Unit 2639 [Otway Saw Mills School No. 3601: 1924-1927]

Unit 2737 [Otway Saw Mills School No. 3601: 1927-1929]

VPRS 795: School Building Files 1868-1975

Unit 1927 [Otway Saw Mills School No. 3601: 1904-1912]

Unit 2811 [Tanybryn School No. 3596: 1935-1956]

Unit 2811 [Mt. Sabine State School No. 3596: 1907-1909]

Unit 2812 [Otway Saw Mills School No. 3601: 1913-1928]

VPRS 796: Outward Correspondence – Primary Schools 1868-1938

Unit 761 [Otway Saw Mills School No. 3601: 1908-1918]

VPRS 1472: Forests Commission Victoria correspondence files pre 1900-1920

Unit 3; files: 'Clark and Pearce – Sawmill Areas, Rubicon'; 'A. Sanderson Sawmill'; 'Hayden Bros Sawmill Area' (Barwon Downs)

VPRS 3524/P1: Criminal Trial Brief Registers 1892-1961

Unit 26: 1918

Department of Sustainability and Environment, Victoria

Carver, M. 1921, Forestry in Victoria (1838-1919), typescript in five volumes.

FCV 'Put Away' Files

Box 55; File 29/0474: 'W. R. Henry and Son'

Geelong Historical Records Centre

Ilie, A. C. 1925, Forester's Annual Report, East Otway District, manuscript.

Shire of Colac Council Rate Books 1923-1926.

South Barwon Shire Council Rate Books 1924-1926.

Winchelsea Shire Council Rate Books 1908-1918.

Yaugher (Forrest) Cemetery Burial Register 1903-1978.

State Library of Victoria

Larkins, A. n.d., River and Range: a story of a Victorian river valley, manuscript, Box 3075/1, MS 12377.

Australian National Library, Canberra

Mackie, J. 1980, Against the elements. Sawmilling in the Otways and at Tanjil Bren. A First-hand Account, typescript.

University of Melbourne Library

Candy, P.C. 1969, The Victorian Cadet Movement: An outline history from 1867 to 1969, typescript.

Published sources

Adams, W. H. 1977, *Silcott, Washington: Ethno-archaeology of a Rural American Community*, WSU Laboratory of Anthropology, Reports of Investigations, No. 54.

Adams, W. H. 1991, 'Trade Networks and Interaction Spheres-A View from Silcott', in *Approaches to Material Culture Research for Historical Archaeologists*, eds. G. L. Miller, O. R. Jones, L. A. Ross and T. Majewski, Society for Historical Archaeology, California University of Pennsylvania, California, Pennsylvania, pp.385-98.

Adams, W. H. 2002, 'Machine Cut Nails and Wire Nails: American Production and Use for Dating 19th-Century and Early-20th-Century Sites', *Historical Archaeology* 36 (4): 66-88.

Adams, W. Y. and Adams, E. W. 1991, *Archaeological typology and practical reality: A dialectical approach to artifact classification and sorting*, Cambridge University Press, Cambridge.

Allison, P. M. 1998, 'The Household in Historical Archaeology', *Australasian Historical Archaeology* 16: 16-29.

Anderson, B. 1991, *Imagined Communities: Reflections on the Origin and Spread of Nationalism*, revised edition, Verso, London and New York.

Appadurai, A. 1986, 'Introduction: commodities and the politics of value', in *The social life of things. Commodities in cultural perspective*, ed. A. Appadurai, Cambridge University Press, Cambridge, pp.3-63.

Arensberg, C. M. 1961, 'The Community as Object and as Sample', *American Anthropologist* 63 (2[1]): 241-64.

Arnold, K. 1985, *Collecting Australian Found Bottles, part 1*, Crown Castleton Publishers, Maiden Gully, Vic.

Arnold, K. 1987, *Collecting Australian Found Bottles, part 2*, Crown Castleton Publishers, Maiden Gully, Vic.

Arnold, K. 1990, *A Victorian Thirst*, Crown Castleton Publishers, Maiden Gully, Vic.

Arnold K. 1995, *Grandpa's Store: Identification and Valuation Guide*, Crown Castleton Publishers, Golden Square, Vic.

Barcan, A. 1988, *Two Centuries of Education in New South Wales*, UNSW Press, Kensington, NSW.

Barr, N. and Cary, J. 1992, *Greening A Brown Land: The Australian Search for Sustainable Land Use*, Macmillan, South Melbourne, Vic.

Baugher-Perlin, S. 1982, 'Analyzing Glass Bottles for Chronology, Function, and Trade Networks', in *Archaeology of Urban America. The Search for Pattern and Process*, ed. R. S. Dickens, Jr., Academic Press, New York, pp.259-89.

Baumann, P. 1999, *Collecting Antique Marbles: Identification and Price Guide*, 3rd edition, Krause Publications, Iola, Wisconsin.

Beaudry, M. C. 1989, 'The Lowell Boott Mills Complex and its Housing: Material Expressions of Corporate Ideology', *Historical Archaeology* 23 (1): 19-32.

Beaudry, M. C. 1995, 'Introduction: Ethnography in Retrospect', in *The Written and the Wrought: Complementary Sources in Historical Anthropology: Essays in Honour of James Deetz*, eds. M. E. D'Agostino, E. Prine, E. Casella and M. Winer, Kroeber Anthropological Society Papers No.79, pp.1-16.

Benson, J. S. and Redpath, P. A. 1997, 'The nature of pre-European native vegetation in south-eastern Australia: a critique of Ryan, D. G., Ryan, J. R. and Starr, B. J. (1995) *The Australian Landscape – Observations of Explorers and Early Settlers*', *Cunninghamia* 5 (2): 285-328.

Bilimoria, P. and Gangopadhyay, R. 1988, 'Indians in Victoria', in *The Australian People*, ed. J. Jupp, Angus & Robertson, Sydney, pp.546-7.

Billis, R. V. and A. S. Kenyon 1932, *Pastoral Pioneers of Port Phillip*, Macmillan, Melbourne.

Birmingham, J. 1992, *Wybalenna: The Archaeology of Cultural Accommodation in Nineteenth Century Tasmania*, Australian Society for Historical Archaeology, Sydney.

Birmingham, J., Jack, I. and Jeans, D. 1983, *Industrial Archaeology in Australia: Rural Industry*, Heinemann, Richmond, Vic.

Blainey, G. 1966, *The Tyranny of Distance*, Sun Books, Melbourne.

Blake, L. J. 1973, *Vision and Realisation. A Centenary History of State Education in Victoria*, 3 volumes, Education Department of Victoria, Melbourne.

Boehm, E. A. 1993, *Twentieth Century Economic Development in Australia*, Longman Cheshire, Melbourne.

Boessneck, J. 1969, 'Osteological differences between Sheep (*Ovis aries* Linne) and Goat (*Capra hircus* Linne)', in *Science in Archaeology*, eds. D. Brothwell and E. Higgs, Thames and Hudson, London, pp.331-58.

Bolton, G. 1992, *Spoils and Spoilers: A history of Australians shaping their environment*, 2nd edition, Allen & Unwin, North Sydney.

Bongiorno, F. 1996, *The People's Party. Victorian Labour and the Radical Tradition 1875-1914*, Melbourne University Press, Melbourne.

Bonyhady, T. 2000, *The Colonial Earth*, Miegunyah at Melbourne University Press, Melbourne.

Boow, J. 1991, *Early Australian Commercial Glass: Manufacturing Processes*, The Heritage Council of New South Wales.

Borschmann, G. 1999, *The People's Forest. A Living History of the Australian Bush*, The People's Forest Press, Blackheath, NSW.

Bourdieu, P. 1984, *Distinction. A Social Critique of the Judgement of Taste*, Routledge & Kegan Paul, London.

Bowman, D. M. J. S. 2000, *Australian Rainforests: Islands of Green in a Land of Fire*, Cambridge University Press, Cambridge.

Branditt, C. F. 1921, 'His Love Story', *Forest, Lake, and Plain* 1 (1): 12-15.

Branditt, C. F. 1922, 'How We Bought Our Library', *Forest, Lake, and Plain* 2 (1): 34.

Brashler, J. G. 1991, 'When Daddy was a Shanty Boy: The Role of Gender in the Organization of the Logging Industry in Highland West Virginia', *Historical Archaeology* 25 (4): 54-68.

Brinkman, R. and Farrell, S. 1990, *Statement of Resources, Uses and Values for the Otway Forest Management Area*, Department of Conservation and Environment, Melbourne.

British Medical Association 1909, *Secret Remedies: What they cost and what they contain*, British Medical Association, London.

Brooks, A. 2005, *An Archaeological guide to British ceramics in Australia 1788-1901*, The Australasian Society for Historical Archaeology and The La Trobe University Archaeology Program, Melbourne.

Broome, R. 1980, *Treasure in Earthen Vessels. Protestant Christianity in New South Wales Society 1900-1914*, University of Queensland Press, St. Lucia, Qld.

Broome, R. 1984, *The Victorians: Arriving*, Fairfax, Syme & Weldon Associates, NSW.

Brough Smyth 1869, *The Gold Fields and Mineral Districts of Victoria*, John Ferres, Government Printer, Melbourne, 1980 facsimile edition, Queensberry Hill Press, Carlton, Vic.

Brown, C. 1993, *The Otways That I Knew*, Merimbula, NSW.

Bunbury, B. 1997, *Timber For Gold: Life on the Goldfields Woodlines 1899-1965*, Fremantle Arts Centre Press, South Fremantle, W. A.

Busch, J. 1991, 'Second Time Around: A Look at Bottle Reuse', in *Approaches to Material Culture Research for Historical Archaeologists*, eds. G. L. Miller, O. R. Jones, L. A. Ross and T. Majewski, Society for Historical Archaeology, California University of Pennsylvania, California, Pennsylvania, pp.113-26.

Calver, M. (ed) 2005, *A Forest Conscienceness*, Millpress, Rotterdam.

Cameron, E. 1986, *Encyclopedia of Pottery and Porcelain. The 19th and 20th Centuries*, faber and faber, London.

Campion, E. 1982, *Rockchoppers. Growing up Catholic in Australia*, Penguin Books, Ringwood, Vic.

Capon, J. 1993, 'Decorative Plasterwork in New South Wales, 1800-1939', *Australasian Historical Archaeology* 11: 43-51

Carney, T. 1981, 'The History of Australian Drug Laws: Commercialism to Confusion?', *Monash University Law Review* 7: 165-204.

Carroll, J. 1988, *The Upper Yarra. An Illustrated History*, Shire of Upper Yarra, Victoria.

Carroll, L. 1999, 'Communities and Other Social Actors: Rethinking Commodities and Consumption in Global Historical Archaeology', *International Journal of Historical Archaeology* 3(3): 31-6.

Carron, L. T. 1985, *A History of Forestry in Australia*, Australian National University Press, Canberra.

Carskadden, J. and Gartley, R. 1990, 'A Preliminary Seriation of 19th-Century Decorated Porcelain Marbles', *Historical Archaeology* 24 (2): 55-69.

Carter, B. 1986, *Gunnersens in Australia. A Century in Timber*, Gunnersens Pty Ltd., Port Melbourne.

Chaucer, G. 1894-97, *The Complete Works of Geoffrey Chaucer*, ed. W. W. Skeat, Clarendon Press, Oxford.

Clark, I. D. 1990, *Aboriginal Languages and Clans: An Historical Atlas of Western and Central Victoria, 1800-1900*, Monash Publications in Geography No. 37, Monash University, Melbourne.

Cohen, A. P. 1985, *The Symbolic Construction of Community*, Ellis Horwood, Chichester.

Coman, B. 1999, *Tooth and Nail: The Story of the Rabbit in Australia*, Text Publishing Company, Melbourne.

Conlin, J. R. 1979, 'Old boy, did you get enough of pie?', *Journal of Forest History*, October 1979: 164-85.

Cook, L. J., Yamin, R. and McCarthy, J. P. 1996, 'Shopping as Meaningful Action: Toward a Redefinition of Consumption in Historical Archaeology', *Historical Archaeology* 30 (4): 50-65.

Cowan, R. S. 1983, *More Work for Mother. The Ironies of Household Technology from the Open Hearth to the Microwave*, Basic Books, New York.

Crook, P., Lawrence, S. and Gibbs, M. 2002, 'The role of artefact catalogues in Australian historical archaeology: a framework for discussion', *Australasian Historical Archaeology* 20: 26-38.

Cubit, S. 1996, *Recollections From the Forest: 75 Years of forest service in Tasmania*, Forestry Tasmania, Hobart.

Cuffley, P. 1984, *Chandeliers and Billy Tea. A Catalogue of Australian Life 1880-1940*, The Five Mile Press, Hawthorn, Vic.

Cumpston, J. H. L. 1989, *Health and Disease in Australia: A History*, Australian Government Publishing Service, Canberra.

Cushion, J. P. 1980, *Handbook of Pottery and Porcelain Marks*, Faber, London.

Cusick, J. G. 1995, 'The Importance of the Community Study Approach in Historical Archaeology, with an Example from Late Colonial St. Augustine', *Historical Archaeology* 29 (4): 59-83.

Dargavel, J. (ed.) 1988, *Sawing, Selling and Sons: Histories of Australian Timber Firms*, Centre for Resource and Environmental Studies, Australian National University, Canberra.

Dargavel, J. 1995, *Fashioning Australia's Forests*, Oxford University Press, Melbourne.

Dargavel, J. and Boutland, P. 1988, 'Establishing Australian sawmilling', in *Sawing, Selling and Sons: Histories of Australian Timber Firms*, ed. J. Dargavel, Centre for Resource and Environmental Studies, Australian National University, Canberra, pp.37-42.

Dargavel, D. Gaughwin, D. and Libbis, B. (eds) 2002, *Australia's Ever-changing Forests V*, Centre for Resource and Environmental Studies, The Australian National University in association with Australian Forest History Society Inc., Canberra.

Dargavel, J. and Libbis, B. (eds) 1999, *Australia's Ever-changing Forests IV*, Centre for Resource and Environmental Studies, The Australian National University in association with Australian Forest History Society Inc., Canberra.

Davies, M. and K. Buckley, 1987, *Archaeological Procedures Manual. Port Arthur Conservation and Development Project*, Occasional Paper no.13, Department of Lands, Parks and Wildlife, Hobart.

Davies, P. 1999, 'Henry's Mill: the archaeology and history of an Otways sawmill settlement', in *Australia's Ever-changing Forests IV*, eds. J. Dargavel and B. Libbis, Centre for Resource and Environmental Studies, The Australian National University in association with Australian Forest History Society Inc., Canberra, pp.247-59.

Davies, P. 2001a, Isolation and Integration: The Archaeology and History of an Otways Forest Community, PhD thesis, La Trobe University, Melbourne.

Davies, P. 2001b, 'A Cure for All Seasons: Health and Medicine in a Bush Community', *Journal of Australian Studies* 70: 63-72.

Davies, P. 2005a, '"The Children are running wild...": Uncovering childhood at a forest sawmill camp', in *A Forest Conscienceness*, ed. M. Calver et al., Millpress, Rotterdam, pp.75-83.

Davies, P. 2005b, 'Space and Structure at an Australian Timber Camp', *Historical Archaeology* 39 (4): 59-72.

Davies, P. 2005c, 'Writing slates and schooling in Victoria', *Australasian Historical Archaeology* 23: 63-69

Davison, G. 1978, *The Rise and Fall of Marvellous Melbourne*, Melbourne University Press, Melbourne.

Dawson, J. 1981 [1881], *Australian Aborigines: The languages and customs of several tribes of Aborigines in the western district of Victoria, Australia*, originally published by George Robertson, Melbourne, facsimile edition by Australian Institute of Aboriginal Studies, Canberra.

Day, D. 1999, *John Curtin: a life*, HarperCollins, Sydney.

Deagan, K. 1982, 'Avenues of Inquiry in Historical Archaeology', *Advances in Archaeological Method and Theory* 5: 151-77.

Deagan, K. 1983, *Spanish St. Augustine: The Archaeology of a Colonial Creole Community*, Academic Press, New York.

Deagan, K. A. 1988, 'Neither History Nor Prehistory; the Questions that Count in Historical Archaeology', *Historical Archaeology* 22 (1): 7-12.

Deetz, J. 1977, *In Small Things Forgotten: The Archaeology of Early American Life*, Anchor Books, New York.

Delle, J. 1998, *An Archaeology of Social Space: Analyzing Coffee Plantations in Jamaica's Blue Mountains*, Plenum Press, New York and London.

Dempsey, K. 1990, *Smalltown. A Study of Social Inequality, Cohesion and Belonging*, Oxford University Press, Melbourne.

Deutsher, K. M. 1999, *The Breweries of Australia: A History*, Lothian Books, Port Melbourne, Vic.

Dingle, T. 1984, *The Victorians: Settling*, Fairfax, Syme & Weldon Associates, NSW.

Doughty, R. W. 2000, *The eucalyptus: a natural and commercial history of the gum tree*, Johns Hopkins University Press, Baltimore.

Douglas, J. G. 1977, 'The Geology of the Otway Region, Southern Victoria', *Proceedings of the Royal Society of Victoria* 89 (part 1): 19-25.

Douglas, M. and Isherwood, B. 1978, *The World of Goods. Towards an Anthropology of Consumption*, Penguin, Ringwood, Vic.

Dovers, S. (ed.) 1994, *Australian Environmental History: Essays and Cases*, Oxford University Press, Melbourne.

Dovers, S. (ed.) 2000, *Environmental history and policy: still settling Australia*, Oxford University Press, Melbourne.

Dummer, G. W. A. 1978, *Electronic Inventions and Discoveries*, Pergamon Press, Oxford.

Elias, N. 1974, 'Towards a Theory of Communities', in *The Sociology of Community*, eds. C. Bell and H. Newby, Frank Cass & Co, London, pp.ix-xli.

Ellis, A. 2001, Toy Stories: Interpreting Childhood from the Victorian Archaeological Record', BA (Honours) thesis, La Trobe University, Melbourne.

Evans, P. 1993, *A Report to the Australian Heritage Commission on Sawmill and Tramway Sites in the Central Highland Forests of Victoria*, 3 volumes, Department of Conservation and Natural Resources, Melbourne.

Evans, P. 1994, *Rails to Rubicon-A history of the Rubicon Forest*, Light Railway Research Society of Australia, Melbourne.

Evans, P. 1997, A Study of Historic Sawmill and Tramway Sites in the North-East Forests Region Victoria (Excluding Box-Ironbark Area), report prepared for Environment Australia and Natural Resources and Environment, Victoria.

Evans, P. 1998, A Study of Historic Sawmill and Tramway Sites in the Gippsland Forest Region Victoria, report prepared for Environment Australia and Natural Resources and Environment, Victoria.

Facey, G. A. 1977, *Otway Timber Mills and Pioneering Days*, published by the author, Anglesea, Vic.

Facey, G. A. 1981, *70 Years in the Otways*, published by the author, Anglesea, Vic.

Fahey, C. 1987, *Barmah Forest: A history of the Barmah Forest*, Department of Conservation, Forests and Lands, Melbourne.

Fainges, M. 1991, *Antique Dolls of China and Bisque*, Kangaroo Press, Kenthurst, NSW.

Ferguson, L. 1977, 'Historical Archaeology and the Importance of Material Things', in *Historical Archaeology and the Importance of Material Things*, ed. L. Ferguson, The Society for Historical Archaeology, Special Publication Series, no. 2, pp.5-8.

Fike, R. E. 1987, *The Bottle Book. A Comprehensive Guide to Historic, Embossed Medicine Bottles*, Peregrine Smith Books, Salt Lake City.

Finch, L. 1998, 'Caring for Colonial Infants: Parenting on the Frontiers', *Australian Historical Studies* 110: 109-26.

Finch, L. 1999, 'Soothing Syrups and Teething Powders: Regulating Proprietary Drugs in Australia, 1860-1910', *Medical History* 43: 74-94.

Fisher, S. W. 1970, *English Pottery and Porcelain Marks*, W. Foulsham & Co., Slough, Berkshire.

Flannery, T. 1994, *The future eaters*, Reed Books, Melbourne.

Floud, R. 1997, *The People and the British Economy 1830-1914*, Oxford University Press, Oxford.

Forests Commission Victoria 1928, *Handbook of Forestry in Victoria*, Government Printer, Melbourne.

Forests Commission Victoria 1931, 'Twelfth Annual report. Financial Year 1930-31', *Papers Presented to Parliament (Victoria)* 1931, vol. 1, no. 18.

Foy and Gibson 1900, *Catalogue of Foy and Gibson*, Foy and Gibson, Collingwood, Vic.

Franzen, J. G. 1992, 'Northern Michigan Logging Camps: Material Culture and Worker Adaptation on the Industrial Frontier', *Historical Archaeology* 26 (2): 74-98.

Frawley, K. J. 1993, 'Logging Technology and Forest Cutting Practices', in *Australia's Ever-Changing*

Forests II, eds. J. Dargavel and S. Feary, Centre for Resource and Environmental Studies, The Australian National University, Canberra, pp.143-63.

Fulekar, M. H. and M. M. Alam Khan 1995, 'Occupational exposure to dust in slate pencil manufacture', *The Annals of Occupational Hygiene* 39(1): 107-14.

Gaw, L. P. 1975, 'The Availability and Selection of Ceramics in Silcott, Washington, 1900-1930', *Northwest Anthropological Research Notes* 1: 166-79.

Geismar, J. H. 1982, *The Archaeology of Social Disintegration in Skunk Hollow: A Nineteenth-Century Rural Black Community*, Academic Press, New York.

Gemmell, W. 1986, *And So We Graft from Six to Six: The Brickmakers of New South Wales*, Angus and Robertson, NSW.

Getty, R. 1975, *Sisson and Grossman's: The Anatomy of the Domestic Animals*, fifth edition, vol. 1, W. B. Saunders, Philadelphia.

Glassie, H. 1982, *Passing the Time. Folklore and History of an Ulster Community*, The O'Brien Press, Dublin.

Glennie, P. 1995, 'Consumption within Historical Studies', in *Acknowledging Consumption. A Review of New Studies*, ed. D. Miller, Routledge, London, pp.164-203.

Godden, G. A. 1991 [1964], *Encyclopedia of British Pottery and Porcelain Marks*, revised edition, Barrie & Jenkins, London.

Gollan, A. 1978, *The Tradition of Australian Cooking*, Australian National University Press, Canberra.

Graeme-Evans, A. 1996, *Against the Odds. Risbys: Tasmanian Timber Pioneers 1826-1995*, Tasbook, Hobart.

Green, D. and Cromwell, L. 1984, *Mutual Aid or Welfare State? Australia's Friendly Societies*, Allen & Unwin, Sydney.

Griffiths, T. 1992, *Secrets of the Forest*, Allen & Unwin, Sydney.

Griffiths, T. 1996, *Hunters and Collectors. The Antiquarian Imagination in Australia*, Cambridge University Press, Cambridge.

Griffiths, T. 2001, *Forests of Ash: An Environmental History*, Cambridge University Press, Cambridge.

Griffiths, T. and Robin, L. (eds.) 1997, *Ecology and Empire: Environmental History of Settler Societies*, Melbourne University Press, Melbourne.

Grimshaw, P., Lake, M., McGrath, A. and Quartly, M. 1994, *Creating A Nation*, McPhee Gribble, South Yarra, Vic.

Grimwade, G. 1998, 'The Canning Stock Route: Desert stock route to outback tourism', *Australasian Historical Archaeology* 16: 70-79.

Haggar, J. 1979, *Australian colonial medicine*, Rigby, Adelaide.

Hall, N., Johnston, R. D. and Chippendale, G. M. 1975, *Forest Trees of Australia*, Australian Government Publishing Service, Canberra.

Halstead, P. and Collins, P. 1994, The Taxonomic Identification of Limb Bone of European Farmyard Animals and Deer: a Multimedia Tutorial, Teaching and Learning Technology Programme, Glasgow.

Hardesty, D. L. 1988, *The Archaeology of Mining and Miners: A View from the Silver State*, The Society for Historical Archaeology, Special Publication Series no. 6, Michigan.

Hardesty, D. L. 2002, 'Commentary: Interpreting Variability and Change in Western Work Camps', *Historical Archaeology* 36(3): 94-8.

Harnwell and Son c. 1920, *Catalogue of One-Fire Stoves, Colonial Ovens, Copper Frames, Etc.*, North Carlton, Melbourne.

Harvey, C. 1992, 'Comet sawmills', in *Secrets of the Forest*, ed. T. Griffiths, Allen & Unwin, Sydney, pp.144-7.

Hay, A. 2002, *Gum*, Duffy & Snellgrove, Potts Point, NSW.

Hebb, I. 1970, *The History of Colac and District*, The Hawthorn Press, Melbourne.

Hechtlinger, A. 1970, *The Great Patent Medicine Era*, Galahad Books, New York.

Hicks, N. 1978, *'This Sin and Scandal': Australia's Population Debate 1891-1911*, Australian National University Press, Canberra.

Higgins, H. B. 1920, Australian Timber Workers' Union vs John Sharpe and Co, *Commonwealth Arbitration Reports, Vol.14*, Commonwealth Court of Conciliation and Arbitration, ed. J. G. B. Castieu, The Law Book Company of Australia Limited, Melbourne, pp.831-6.

Hill, S. H. 1982, 'An Examination of Manufacture-Deposition Lag for Glass Bottles from Late Historic Sites', in *Archaeology of Urban America. The Search for Pattern and Process*, ed. R. S. Dickens, Jr., Academic Press, New York, London, pp.291-327.

Hillery, G. A., Jr. 1955, 'Definitions of Community: Areas of Agreement', *Rural Sociology* 20 (2): 111-123.

Hirst, J. B. 1975, 'Distance in Australia-Was it a Tyrant?' *Historical Studies* 16: 435-47.

Hobsbawm, E. 1979, *The Age of Capital 1848-1875*, The New American Library, New York.

Holt, C. A. 1991, 'Plants, Humans, and Culture: An Edible Model of Consuming Behavior', *Historical Archaeology* 25 (2): 46-61.

Hooker, C. 2000, 'Diphtheria, Immunisation and the Bundaberg tragedy: A study of Public Health in Australia', *Health and History* 2 (1): 52-78.

Hordern, A. and Sons 1897, *General Catalogue*, Anthony Hordern, Haymarket, NSW.

Houghton, N. 1975, *Sawdust and Steam. A history of the railways and tramways of the eastern Otway ranges*, Light Railway Research Society of Australia, Melbourne.

Houghton, N. 1976, *Forrest: The First 85 Years*, Light Railway Research Society of Australia, Vic.

Houghton, N. 1979, *Barongarook: A Rural Portrait*, Barongarook School Centenary Committee, Vic.

Houghton, N. 1980, *Timber and Gold: A history of the sawmills and tramways of the Wombat Forest 1855-1940*, Light Railway Research Society of Australia, Melbourne.

Houghton, N. 1982, 'Victoria's Pioneer Sawmill and Timber Tramway', *Light Railways* 75: 51-3.

Houghton, N. 1986, *Timber Mountain. A Sawmilling History of the Murrindindi Forest 1885-1950*, Light Railway Research Society of Australia, Melbourne.

Houghton, N. 1988, 'Rail or Road? The State's Policy Dilemma on Timber Transport in the Otway Ranges 1890-1955', in *Australia's Ever Changing Forests*, eds. K. J. Frawley and N. M. Semple, Special Publication No. 1, Department of Geography and Oceanography, Australian Defence Force Academy, Campbell, ACT, pp.291-9.

Houghton, N. 1992, *The Beechy. The life and times of the Colac-Beech Forest-Crowes narrow gauge railway 1902-1962*, Light Railway Research Society of Australia, Melbourne.

Houghton, N. 1995, 'Henry's Locos, Forrest, Victoria', *Light Railways*, January 1995, pp.6-11.

Houghton, N. 1997, 'East Otway odyssey of rugged field work', in *Australia's Ever-Changing Forests III*, ed. J. Dargavel, Centre for Resource and Environmental Studies, The Australian National University, Canberra, pp.353-60.

Houghton, N. forthcoming, Sawdust and Steam: A Sawmilling History of the East Otway Ranges.

Howitt, W. 1855, *Land, Labour and Gold, or, Two Years in Victoria with Visits to Sydney and Van Diemen's Land*, Lowden, Kilmore.

Hutchison, D. 1987, 'Identifying Bottles', in *Papers in Australian Historical Archaeology*, eds. J. Birmingham and D. Bairstow, Australian Society for Historical Archaeology, Sydney, pp.153-60.

Hutton, B. 1988, 'The first forest sawmills at Mount Macedon', in *Sawing, Selling and Sons: Histories of Australian Timber Firms*, ed. J. Dargavel, Centre for Resource and Environmental Studies, Australian National University, Canberra, pp.31-3.

Iacono, N. Z. 1999, 'Miscellaneous Artefacts Reports', *The Cumberland/Gloucester Streets Site, The Rocks. Archaeological Investigation Report*, vol. 4, part II, Godden Mackay Logan Pty Ltd Heritage Consultants, Redfern NSW, pp.11-118.

Isaacs, J. 1990, *Pioneer Women of the Bush and Outback*, Lansdowne Press, Sydney.

Ivey, W. E. 1874, 'Papers relating to forest conservancy', *Papers Presented to Parliament (Victoria)*, vol. 3, no. 86.

Jack, I., Holmes, K. and Kerr, R. 1984, 'Ah Toy's Garden: A Chinese Market-Garden on the Palmer River Goldfield, North Queensland', *The Australian Journal of Historical Archaeology* 2: 51-8.

Jack, R. I. and Cremin, A. 1994, *Australia's Age of Iron: History and Archaeology*, Oxford/Sydney University Press, South Melbourne, Vic.

Jackson Lears, T. J. 1983, 'From Salvation to Self-Realization: Advertising and the Therapeutic Roots of the Consumer Culture, 1880-1930', in *The Culture of Consumption: Critical Essays in American History*, eds. R. W. Fox and T. J. Jackson Lears, Pantheon Books, New York, pp.3-38.

Jardine, L. 1996, *Worldly Goods*, Macmillan, London.

Jenkins, E. *c.* 1890, *Catalogue of bedstead, range, grate, fender, one-fire stove, café, table, and hall furniture manufacturer, general iron founder*, Melbourne.

Johnson, A.W. 1999, 'Coins, Medals and Tokens Artefact Report', in *The Cumberland/Gloucester Streets Site,*

The Rocks. Archaeological Investigation Report, vol. 4 part 2, Godden Mackay Logan, Sydney, pp.239-84.

Johnson, M. H. 1999, 'Rethinking historical archaeology', in *Historical Archaeology. Back from the edge*, eds. P. P. A. Furani, M. Hall and S. Jones, Routledge, London and New York, pp.23-36.

Jones, F. 1976, *Not by myself. The Fletcher Jones Story*, The Wentworth Press, Surry Hills, NSW.

Jones, O. 1981, 'Essence of Peppermint. A History of the Medicine and its Bottle', *Historical Archaeology* 15 (2): 1-57.

Jones, O. 2000, 'A Guide to Dating Glass Tableware: 1800 to 1940', in *Studies in Material Culture Research*, ed. K. Karklins, The Society for Historical Archaeology, Anthropology Section, California University of Pennsylvania, California, Pennsylvania, pp.141-232.

Jones, O. and Sullivan, C. 1985, *The Parks Canada Glass Glossary for the Description of Containers, Tableware, Flat Glass and Closures*, National Historic Parks and Sites, Canadian Parks Service, Environment Canada, Ottawa.

Jones, P. (ed) 1981, *Historical Records of Victoria, Volume 1: Beginnings of Permanent Government*, Victorian Government Printing Office, Melbourne.

Karskens, G. 1999, *Inside the Rocks. The Archaeology of a Neighbourhood*, Hale & Iremonger, Alexandria, NSW.

Karskens, G. and Thorpe, W. 1992, 'History and Archaeology in Sydney: Towards Integration and Interpretation', *Journal of the Royal Australian Historical Society* 78 (3 and 4): 52-75.

Keller, S. 2003, *Community: Pursuing the Dream, Living the Reality*, Princeton University Press, Princeton and Oxford.

Kingston, B. 1994, *Basket, Bag and Trolley. A history of Shopping in Australia*, Oxford University Press, Melbourne.

Kostoglou, P. 1992, *Mount Horror State Forest: An Archaeological Survey of the Historic Timber Industry*, Archaeology of the Tasmanian Timber Industry Report No. 2, Forestry Commission, Tasmania.

Kostoglou, P. 1994, *Historic timber-getting between Hastings and Dover Block 2*, Archaeology of the Tasmanian Timber Industry Report No. 5, Forestry Tasmania and Tasmanian Forest Research Council, Hobart.

Lake, M. 1986, 'The Politics of Respectability: Identifying the Masculinist Context', *Historical Studies* 22: 116-31.

Lake, M. 1987, *The Limits of Hope. Soldier Settlement in Victoria 1915-38*, Oxford University Press, Melbourne.

Lake, M. 1991, 'Historical Homes', *Australian Historical Studies* 24: 46-54.

Land Conservation Council, Victoria 1976, *Report on the Corangamite Study Area*, Melbourne.

Landon, C. and Tonkin, D. 1999, *Jackson's Track. Memoir of a Dreamtime Place*, Viking, Ringwood, Vic.

Lassetter, F. and Co. 1902, *Lassetters' Commercial Review*, Sydney.

Laughton, A.M. 1915-16, *Victorian Year Book* 1915-16, Government Printer, Melbourne.

Lawrence Cheney, S. 1995, No Abiding City: The Archaeology and History of an Ephemeral Mining Settlement, PhD thesis, School of Archaeology, La Trobe University, Vic.

Lawrence, S. 2000, *Dolly's Creek. An Archaeology of a Victorian Goldfields Community*, Melbourne University Press, Carlton South, Victoria.

Lawrence, S. 2003a, 'Exporting Culture: Archaeology and the Nineteenth-Century British Empire', *Historical Archaeology* 37 (1): 20-33.

Lawrence, S. 2003b, 'At home in the bush: material culture and Australian nationalism', in *Archaeologies of the British: Explorations of identity in Great Britain and its colonies 1600-1945*, ed. S. Lawrence, Routledge, London, pp.211-23.

Lemon, A. and M. Morgan 1986, *Poor Souls, They Perished: The* Cataraqui, *Australia's Worst Shipwreck*, Hargreen Publishing, North Melbourne, Vic.

Lennon, J. 1988, 'Timeless Wilderness? The Use of Historical Source Material in Understanding Environmental Change in Gippsland, Victoria', in *Australia's Ever Changing Forests*, eds. K. J. Frawley and N. M. Semple, Special Publication No. 1, Department of Geography and Oceanography, Australian Defence Force Academy, Campbell, ACT, pp.419-40.

Lennon, J. 1998, 'History, Cultural Heritage and the Regional Forest Assessment Process', *Australian Journal of Environmental Management* 5: 38-46.

Leone, M. P. 1988, 'The Relationship Between Archaeological Data and the Documentary Record: 18th Century Gardens in Annapolis, Maryland', *Historical Archaeology* 22 (1): 29-35.

Leone, M. P. 1999, 'Setting Some Terms for Historical Archaeologies of Capitalism', in *Historical Archaeologies of Capitalism*, eds. M. P. Leone and P. Parker Potter, Jr., Kluwer Academic/Plenum Publishers, New York, pp.3-20.

Leone, M. P. and Crosby, C. A. 1987, 'Epilogue: Middle-Range Theory in Historical Archaeology', in *Consumer Choice in Historical Archaeology*, ed. S. M. Spencer-Wood, Plenum Press, New York and London, pp.397-410.

Leone, M. P. and Potter, P. Parker, Jr. 1988, 'Introduction: Issues in Historical Archaeology', in *The Recovery of Meaning. Historical Archaeology in the Eastern United States*, eds. M. P. Leone and P. Parker Potter, Jr., Smithsonian Institution Press, Washington and London, pp.1-22.

Lewis, K. E. 1984, *The American Frontier: An Archaeological Study of Settlement Pattern and Process*, Academic Press, Orlando, Florida.

Lewis, M. 1992, *A Rum State. Alcohol and State Policy in Australia 1788-1988*, Australian Government Publishing Service, Canberra.

Ligar, C. W., Hodgkinson, C. and Smyth, R. B. 1864-65, 'Report on the advisableness of establishing State forests', *Papers Presented to Parliament (Victoria)*, vol. 4, part 2, no. 77.

Lindbergh, J. 1999, 'Buttoning Down Archaeology', *Australasian Historical Archaeology* 17: 50-7.

Lindsay, J. 1974, *A History of the North Wales Slate Industry*, David and Charles, Newton Abbot, Devon.

Linforth, D. J. 1977, 'The Climate of the Otway Region', *Proceedings of the Royal Society of Victoria* 89 (part 1): 61-8.

Linge, G. J. R. 1979, *Industrial Awakening. A Geography of Australian Manufacturing 1788 to 1890*, Australian National University Press, Canberra.

Little, B. J. 1992, 'Text-Aided Archaeology', in *Text-Aided Archaeology*, ed. B. J. Little, CRC Press, Boca Raton, Florida, pp.1-6.

Little, B. J. 1994, 'People with History: An Update on Historical Archaeology in the United States', *Journal of Archaeological Method and Theory* 1 (1): 5-40.

Loney, J. 1979, *Otway Memories*, Marine History Oublications, Geelong, Vic.

Loney, J. 1989, *Otway and its coast in retrospect*, Marine History, Portarlington, Vic.

Lonie, J. 1979, *A Social History of Drug Control in Australia*, Research paper 8, Royal Commission into the Non-Medical Use of Drugs, South Australia.

Lorrain, D. 1968, 'An Archaeologist's Guide to Nineteenth Century American Glass', *Historical Archaeology* 2: 35-44.

Lowenthal, D. 1985, *The Past is a Foreign Country*, Cambridge University Press, Cambridge.

Lydon, J. 1999, *Many Inventions. The Chinese in the Rocks, Sydney, 1890-1930*, Monash Publications in History, Melbourne.

Mackay, H. 1908, 'Report of the State Forests Department of Victoria', *Papers Presented to Parliament (Victoria)* 1908 vol. II, no. 41.

Mackay, H. 1909, 'Report of the State Forests Department of Victoria, 1908-9', *Papers Presented to Parliament (Victoria)* 1909 vol. II (2nd session), no. 37.

Mackay, H. 1914, 'Report of the State Forests Department of Victoria, 1913-14', *Papers Presented to Parliament (Victoria)* 1914 (2nd session), no. 4.

Majewski, T. and O'Brien, M. J. 1987, 'The Use and Misuse of Nineteenth-Century English Ceramics in Archaeological Analysis', in *Advances in Archaeological Method and Theory* 11: 97-209.

Manderson, D. 1993, *From Mr Sin to Mr Big. A History of Australian Drug Laws*, Oxford University Press, Melbourne.

Marsden, S. 1993, 'What's Wrong with the Twentieth Century? Why Has it Been Ignored by Archaeologists?' *Australasian Historical Archaeology* 11: 142-4.

Martyr, P. 2002, *Paradise of Quacks: an Alternative History of Medicine in Australia*, Macleay Press, Sydney.

Matthews, K. 1999, 'Familiarity and Contempt. The archaeology of the "modern"', in *The Familiar Past? Archaeologies of later historical Britain*, eds. S. Tarlow and S. West, Routledge, New York and London, pp.155-79.

Mayne, A. and Lawrence, S. 1998, 'An Ethnography of Place: Imagining "Little Lon"', *Journal of Australian Studies* 57: 93-107.

McCalman, J. 1984, *Struggletown. Portrait of an Australian Working-Class Community*, Penguin Books, Ringwood, Vic.

McCarthy, M. J. 1983, *Trestle Bridges and Tramways: The Timber Industry of Erica District 1910-1950*, Light Railway Research Society of Australia, Melbourne.

McCarthy, M. J. 1987, *Bellbrakes, Bullocks and Bushmen. A Sawmilling and Tramway History of Gembrook 1885-1985*, Light Railway Research Society of Australia, Melbourne.

McCarthy, M. J. 1993, *Settlers and Sawmillers: A History of West Gippsland Tramways and the Industries they served, 1875-1934*, Light Railway Research Society of Australia, Melbourne.

McCarthy, M. J. 2001, *Mountains of Ash: A History of the Sawmills and Tramways of Warburton and District*, Light Railway Research Society of Australia, Melbourne.

McCarthy, P. G. 1976, 'Justice Higgins and the Harvester Judgement', in *Social Policy in Australia: Some Perspectives*, ed. J. Roe, Casell Australia, Stanmore, NSW, pp.41-59.

McCracken, G. 1988, *Culture and Consumption. New Approaches to the Symbolic Character of Consumer Goods and Activities*, Indiana University Press, Bloomington.

McFarlane, I. (ed) 1982, 'The Wesleyan Mission at Buntingdale' in *Historical Records of Victoria: Volume 2A: The Aborigines of Port Phillip 1835-1839*, Victorian Government Printing Office, Melbourne, pp.75-150.

McIntosh, I. 1988, *Forest, Lake and Plain: The History of Colac 1888-1988*, Hargreen Publishing, Vic.

McKendrick, N., Brewer, J. and Plumb, J. H. 1982, *The Birth of a Consumer Society. The Commercialization of Eighteenth-Century England*, Indiana University Press, Bloomington.

McRae, H. 1994, *Forest History in Victoria. A Guide to Government Records 1836-1994*, Historic Places Section, Department of Conservation and Natural Resources, Victoria.

Miller, D. 1987, *Material Culture and Mass Consumption*, Basil Blackwell, Oxford.

Miller, D. 1995, 'Consumption as the Vanguard of History: A Polemic by Way of an Introduction', in *Acknowledging Consumption. A Review of New Studies*, ed. D. Miller, Routledge, London, pp.1-57.

Miller, G. L. 1991a, 'Classification and Economic Scaling of 19th Century Ceramics', in *Approaches to Material Culture Research for Historical Archaeologists*, eds. G. L. Miller, O. R. Jones, L. A. Ross and T. Majewski, Society for Historical Archaeology, California University of Pennsylvania, California, Pennsylvania, pp.37-58.

Miller, G. L. 1991b, 'A Revised Set of CC Index Values for Classification and Economic Scaling of English Ceramics from 1787 to 1880', *Historical Archaeology* 25: 1-25.

Miller, G. L. 1992, 'Thoughts Towards a Users' Guide to Ceramic Assemblages, Part Three', *Council for Northeast Historical Archaeology Newsletter* 22: 2-4.

Miller, G. L. 1993, 'A Users' Guide to Ceramic Assemblages: Part Four', *Council for Northeast Historical Archaeology Newsletter* 26: 4-7.

Miller, G. L. and Sullivan, C. 1984, 'Machine-Made Containers and the End of Production for Mouth-Blown Bottles', *Historical Archaeology* 18: 83-96.

Mills, J. 1986, *The Timber People. A History of Bunnings Limited*, Bunnings Limited, Perth.

Minutes of Evidence 1899, Royal Commission on State Forests and Timber Reserves, 1897-1901.

Monod, D. 1996, *Store Wars: shopkeepers and the culture of mass marketing, 1890-1939*, University of Toronto Press, Toronto.

Moulds, F. R. 1991, *The Dynamic Forest*, Lynedoch Publications, Melbourne.

Mrozowski, S. A., Ziesing, G. H. and Beaudry, M. C. 1996, *Living on the Boott: historical archaeology at the Boott Mills Boardinghouses, Lowell, Massachusetts*, University of Massachusetts Press, Amherst.

Mueller, F. von 1871, *Forest Culture in its Relation to Industrial Pursuits*, Government Printer, Melbourne.

Mukerji, C. 1983, *From Graven Images: Patterns of Modern Materialism*, Columbia University Press, New York.

Murray, T. and A. Mayne 2001, 'Imaginary landscapes: reading Melbourne's "Little Lon"', in *The Archaeology of Urban Landscapes: Explorations in Slumland*, eds. A. Mayne and T. Murray, Cambridge University Press, Cambridge, pp.89-105.

Murray, T. and A. Mayne 2003, '(Re)Constructing a Lost Community: "Little Lon", Melbourne, Australia', *Historical Archaeology* 37(1): 87-101.

Nelson, H. 1988, 'Minimay: One of Six Thousand Weatherboard Schools', *Australian Cultural History* 7: 5-17.

O'Farrell, P. 1977, *The Catholic Church and Community in Australia. A History*, Nelson, West Melbourne.

Orser, C. E., Jr. 1999, 'Archaeology and the Challenges of Capitalist Farm Tenancy in America', in *Historical Archaeologies of Capitalism*, eds. M. P. Leone and P. Parker Potter, Jr., Kluwer Academic/Plenum Publishers, New York, pp.143-67.

Parks Canada 1992, *Classification System for Historical Collections*, Canadian Parks Service, Environment Canada, Ottawa.

Parsons, R. F., Kirkpatrick, J. B. and Carr, G. W. 1977, 'Native vegetation of the Otway Region, Victoria', *Proceedings of the Royal Society of Victoria*, vol. 89 (part 1): 77-88.

Parsonson, I. 1998, *The Australian Ark: A History of Domesticated Animals in Australia*, CSIRO Publishing, Melbourne.

Parssinen, T. M. 1983, *Secret Passions, Secret Remedies. Narcotic Drugs in British Society 1820-1930*, Manchester University Press, Manchester.

Pemberton, B. 1979, *Australian Coastal Shipping*, Melbourne University Press, Melbourne.

Pensabene, T. S. 1980, *The rise of the medical practitioner in Victoria*, Health Research Project Research Monograph 2, The Australian National University, Canberra.

Perrin, G. S. 1890, 'Report of the Conservator of Forests for the Year Ending 30th June, 1890', *Papers Presented to Parliament (Victoria)*, vol. IV, no. 202.

Poplin, D. E. 1972, *Communities. A Survey of Theories and Methods of Research*, Macmillan, New York.

Portus, J. H. 1971, *Australian Compulsory Arbitration 1900-1970*, Hicks Smith and Sons, Sydney.

Powell, J. M. (ed) 1973, *Yeoman and Bureaucrats: The Victorian Crown Lands Commission 1878-79*, Oxford University Press, Melbourne.

Powell, J. M. 1976, *Environmental Management in Australia 1788-1914*, Oxford University Press, Melbourne.

Praetzellis, M. and Praetzellis, A. 1990a, *"For a Good Boy": Victorians on Sacramento's J Street*, Cultural Resources Facility, Anthropological Studies Center, Sonoma State University, Rohnert Park, California.

Praetzellis, M. and Praetzellis, A. 1990b, *Junk! Archaeology of the Pioneer Junk Store, 1877-1908*, Papers in Northern California Anthropology no. 4, Anthropological Studies Center, Sonoma State University, Rohnert Park, California.

Praetzellis, A. and Praetzellis, M. 1992, 'Faces and Facades: Victorian Ideology in Early Sacramento', in *The Art and Mystery of Historical Archaeology. Essays in Honour of James Deetz*, eds. A. E. Yentsch and M. C. Beaudry, CRC Press, Boca Raton, pp.75-99.

Pratt, A. 1978, *Sidney Myer: A Biography*, Quartet, Melbourne.

Presland, G. 1982, *An Archaeological Survey of the Otway Forest Region*, Victoria Archaeological Survey Occasional Reports Series No. 8, Melbourne.

Priestley, S. 1986, *Bush Nursing in Victoria. 1910-1985. The First 75 Years*, Victorian Bush Nursing Association in conjunction with the Lothian Publishing Company, Melbourne.

Purser, M. 1992, 'Consumption as Communication in Nineteenth-Century Paradise Valley, Nevada', *Historical Archaeology* 26 (3): 105-116.

Purser, M. 1999, '*Ex Occidente Lux*? An Archaeology of Later Capitalism in the Nineteenth-Century West', in *Historical Archaeologies of Capitalism*, eds. M. P. Leone and P. Parker Potter, Jr., Kluwer Academic/ Plenum Publishers, New York, pp.115-41.

Pyne, S. J. 1991, *Burning Bush: A Fire History of Australia*, Henry Holt, New York.

Quiggan, P. 1988, *No Rising Generation: Women and Fertility in late Nineteenth Century Australia*, Australian Family Formation Project Monograph no.10, Department of Demography, Research School of Social Sciences, Australian National University, Canberra.

Radforth, I. 1987, *Bushworkers and Bosses: Logging in Northern Ontario, 1900-1980*, University of Toronto Press, Toronto.

Rait, L. 1989, *Through the Nursery Window: A History of Antique and Collectible Dolls in Australia, 1788-1950*, Oxford University Press, Melbourne.

Rajala, R. A. 1989, 'Bill and the Boss: Labor Protest, Technological Change, and the Transformation of the West Coast Logging Camp, 1890-1930', *Journal of Forest History*, October 1989: 168-79.

Randall, M. E. 1971, 'Early Marbles', *Historical Archaeology* 5: 102-5.

Reiger, K. M. 1985, *The Disenchantment of the Home. Modernizing the Australian family 1880-1940*, Melbourne University Press, Melbourne.

Report 1899, 'Fifth Progress Report of the Royal Commission on State Forests and Timber Reserves. The Otway Forest: Its Resources, Management, and Control', *Papers Presented to Parliament (Victoria)*, vol. III (2), no. 45.

Report 1907, Report of the Royal Commission on Secret Drugs, Cures, and Foods, *Papers Presented to Parliament (Commonwealth)*, Session 1907-8, vol. IV.

Report 1920-1, 'Report of the Royal Commission on the Basic Wage', *Commonwealth Parliamentary Papers 1920-1*, vol. IV.

Report 1926, 'Fourth Report of the Commission of Public Health to the Minister of Public Health', *Papers Presented to Parliament (Victoria)*, vol. II, no. 38.

Ribbentrop, G. S. 1896, 'Report on the State forests of Victoria', *Papers Presented to Parliament (Victoria)*, vol. II, no. 42.

Richards, T. 1990, *The Commodity Culture of Victorian England. Advertising and Spectacle 1851-1914*, Stanford University Press, Stanford, California.

Ridley, J. C. 1997, *Sick, Not Dead: The Health of British Workingmen during the Mortality Decline*, The Johns Hopkins University Press, Baltimore and London.

Riordan, T. B. and Adams, W. H. 1985, 'Commodity Flows and National Market Access', *Historical Archaeology* 19 (2): 5-18.

Roberts, R. 1971, *The Classic Slum. Salford life in the first quarter of the century*, Manchester University Press, Manchester.

Robin, L., T. Sharratt and T. Griffiths (eds) 2005, *A Change in the Weather: Climate and Culture in Australia*, National Museum of Australia Press, Canberra.

Rock, J. T. 1984, 'Cans in the Countryside', *Historical Archaeology* 18 (2): 97-111.

Rogers, N. 1998, *An Alternative Path. The Making and Remaking of Hahnemann Medical College and Hospital of Philadelphia*, Rutgers University Press, New Brunswick.

Room, R. 1988, 'The dialectic of drinking in Australian life: from the Rum Corps to the wine column', *Australian Drug and Alcohol Review* 7: 413-37.

Roseberry, W. 1989, *Anthropologies and Histories: Essays in Culture, History and Political Economy*, Rutgers University Press, New Brunswick and London.

Ross, L. A. and Light, J. D. 2000, 'A Guide to the Description and Interpretation of Metal Files', in *Studies in Material Culture Research*, ed. K. Karklins, The Society for Historical Archaeology, Anthropology Section, California University of Pennsylvania, California, Pennsylvania, pp.20-31.

Rotman, D. L. and Nassaney, M. S. 1997, 'Class, Gender, and the Built Environment: Deriving Social Relations from Cultural Landscapes in Southwest Michigan', *Historical Archaeology* 31 (2): 42-62.

Rutman, D. B. and Rutman, A. H. 1984, *A Place in Time: Middlesex County, Virginia 1650-1750*, W. W. Norton & Co, New York.

Rutterford, K. 1987, *Collecting Shotgun Cartridges*, Stanley Paul, London.

Sands and McDougall's Melbourne, Suburban and Country Directory, 1900 – 1940.

Scales, I. A. 1983, *The timber milling industry at Wye River, Vic.*, Abbotsford, Vic.

Scarlett, N. H. 1977, 'The Aborigines of the Otway Region', *Proceedings of the Royal Society of Victoria* 89 (1): 1-5.

Schama, S. 1995, *Landscape and Memory*, HarperCollins, London.

Schiffer, M. B. 1987, *Formation Processes of the Archaeological Record*, University of New Mexico Press, Albuquerque.

Schlereth, T. J. 1989, 'Country Stores, County Fairs, and Mail-Order Catalogues: Consumption in Rural America', in *Consuming Visions. Accumulation and Display of Goods in America 1880-1920*, ed. S. J. Bronner, W. W. Norton & Company, New York, pp.339-75.

Schlereth, T. J. 1991, *Victorian America: Transformations in everyday life, 1876-1915*, HarperCollins, New York.

Schmid, E. 1972, *Atlas for the Identification of Animal Bones*, Elsevier, Amsterdam.

Schudson, M. 1984, *Advertising, The Uneasy Persuasion: Its Dubious Impact on American Society*, Basic Books, New York.

Schuyler, R. L. 1978a, 'Historical and Historic Sites Archaeology as Anthropology: Basic Definitions and

Relationships', in *Historical Archaeology. A Guide to Substantive and Theoretical Contributions*, ed. R. L. Schuyler, Baywood Publishing Company, New York, pp.27-31.

Schuyler, R. L. 1978b, 'The Spoken Word, the Written Word, Observed Behavior and Preserved Behavior: The Contexts Available to the Archaeologist', in *Historical Archaeology. A Guide to Substantive and Theoretical Contributions*, ed. R. L. Schuyler, Baywood Publishing Company, New York, pp.269-77.

Scott, E. M. 1994, 'Through the Lens of Gender: Archaeology, Inequality, and Those "of Little Note"', in *Those of Little Note: Gender, Race, and Class in Historical Archaeology*, ed. E. M. Scott, University of Arizona Press, Tucson, pp.3-24.

Sears, Roebuck and Co, 1897 Catalogue, Chicago.

Seddon, G. 1994, *Searching for the Snowy: an environmental history*, Allen & Unwin, St. Leonards, NSW.

Seebeck, J. H. 1979, 'G. F. Seebeck: Stonies, Soft Drinks and Sawmills', *Australian Bottle Review* 29: 14-15.

Shackel, P. A. 1993, *Personal Discipline and Material Culture. An Archaeology of Annapolis, Maryland, 1695-1870*, The University of Tennessee Press, Knoxville.

Shammas, C. 1990, *The Pre-industrial Consumer in England and America*, Clarendon Press, Oxford.

Shorter, E. 1986, *Bedside Manners. The Troubled History of Doctors and Patients*, Viking, New York.

Shueard, H. and Tuckwell, D. 1993, *Brewers and aerated water manufacturers in South Australia:1836-1936*, Stepney, S. A.

Sinclair, E. K. 1990, *The Spreading Tree. A History of APM and AMCOR 1844-1989*, Allen & Unwin, Sydney.

South, S. 1977, *Method and Theory in Historical Archaeology*, Academic Press, New York.

Spennemann, D. and Colley, S. 1990, 'Fire in a Pit: The Effects of Burning on Faunal Remains', *Archaeozoologia* 3 (1-2): 45-63.

Sprague, R. 1980-81, 'A Functional Classification for Artifacts from 19th and 20th Century Historical Sites', *North American Archaeologist* 2 (3): 251-61.

Stage, S. 1979, *Female Complaints. Lydia Pinkham and the Business of Women's Medicine*, W. W. Norton & Company, New York.

Stamford, F. E., Stuckey, E. G. and Maynard, G. L. 1984, *Powelltown: A history of its Timber Mills and Tramways*, Light Railway Research Society of Australia, Melbourne.

Starr, P. 1982, *The Social Transformation of American Medicine*, Basic Books, New York.

Stevens, C. 1989, *Tin Mosques and Ghantowns. A History of Afghan Cameldrivers in Australia*, Oxford University Press, Melbourne.

Stewart-Abernathy, L. C. 1992, 'Industrial Goods in the Service of Tradition: Consumption and Cognition on an Ozark Farmstead before the Great War', in *The Art and Mystery of Historical Archaeology. Essays in Honour of James Deetz*, eds. A. E. Yentsch and M. C. Beaudry, CRC Press, Boca Raton, pp.101-26.

Story, A. and Davies, P. 1995, *Historic Forest and Forest-Based Places in South-West Victoria*, report to the Victorian Land Conservation Council.

Strasser, S. 1989, *Satisfaction Guaranteed. The Making of the American Mass Market*, Pantheon Books, New York.

Stratton, M. and B. Trinder 2000, *Twentieth Century Industrial Archaeology*, E & FN Spon, London.

Stretton, L. E. B. 1939, *Report of the Royal Commission to inquire into the causes of and measures taken to prevent the bush fires of January, 1939*, Government Printer, Melbourne.

Stuart, I.-M. F. 1981, 'Ethnohistory in the Otway Ranges', *The Artefact* 6 (1 and 2): 79-88.

Stuart, I. 1989, 'Why Did the Hoffman Brick and Pottery Works Stop Making Bricks?', *The Australian Journal of Historical Archaeology* 7: 29-34.

Stuart, I. 1993, 'Bottles for Jam? An Example of Recycling From a Post-Contact Archaeological Site', *Australian Archaeology* 36: 17-21.

Summers, A. 1994, *Damned Whores and God's Police*, 2nd edition, Penguin Books, Ringwood, Vic.

Sussman, L. 1985, *The Wheat Pattern. An Illustrated Survey*, Studies in Archaeology, Architecture and History, Parks Canada, Environment Canada.

Sussman, L. 1997, *Mocha, Banded, Cat's Eye, and Other Factory-Made Slipware*, Studies in Northeast Historical Archaeology No. 1, Boston University, Boston.

Talbot, O. 1974, 'The Evolution of Glass Bottles for Carbonated Drinks', *Post Medieval Archaeology* 8: 29-62.

Taylor, A. 1998, *A Forester's Log*, Melbourne University Press, Melbourne.

Thearle, M. J. 1990, 'Domestic Medicine in Colonial Queensland', in *New Perspectives on the History of Medicine*, eds. H. Attwood, R. Gillespie and Lewis, University of Melbourne and the Australian Society of the History of Medicine, Parkville, Vic., pp.121-41.

Toulouse, J. H. 1971, *Bottle Makers and their Marks*, Thomas Nelson Inc., New York, Camden.

Tönnies, F. 1963 [1887], *Community and Society*, Harper & Row, New York (translated and edited by C. P. Loomis).

Tregenza, J. 1991, *Le Messuriers of Port Adelaide*, Le Messurier Timber Company Pty Ltd., Port Adelaide.

Tupper, H. 1965, *To the Great Ocean: Siberia and the Trans-Siberian Railway*, Secker & Warburg, London.

Turner, N. 1992, *Catholics in Australia. A Social History*, Collins Dove, North Blackburn, Vic.

Twigg, K. 1994, *'Sixpence for a piece of Timber...' A History of Bowen and Pomeroy 1894-1994*, Bowens, North Mulgrave, Vic.

Tyrrell, I. 1999, *True Gardens of the Gods: Californian-Australian Environmental Reform, 1860-1930*, University of California Press, Berkeley.

Upton, D. and Vlach, J. M. 1986, 'Introduction', in *Common Places: Readings in American Vernacular Architecture*, eds. D. Upton and J. M. Vlach, The University of Georgia Press, Athens and London, pp.xiii-xxiv.

Vader, J. and Murray, M. 1975, *Antique Bottle Collecting in Australia*, Ure Smith, Sydney.

Van Bueren, T. M. 2002, 'The Changing Face of Work in the West: Some Introductory Comments', *Historical Archaeology* 36(3): 1-7.

Varman, R. V. J. 1987, 'The Nail as a Criterion for the Dating of Building and Building Sites (Late 18th Century to 1900)', in *Papers in Australian Historical Archaeology*, eds. J. Birmingham and D. Bairstow, Australian Society for Historical Archaeology, Sydney, pp.104-112.

Vermeulen, P. A. V. R. 1979, 'Dutch Geneva and its bottles', *Australian Bottle Review* 5 (29): 18-19.

Vine Hall, N. 1994, *Tracing Your Family History in Australia*, Albert Park, Vic.

Vines, G. 1985, The Historical Archaeology of Forest Based sawmilling in Victoria 1855-1940, BA (Hons) thesis, La Trobe University, Vic.

Walker, R. 1984, *Under Fire. A History of Tobacco Smoking in Australia*, Melbourne University Press, Melbourne.

Warren, R. L. 1972, *The Community in America*, Rand McNally College Publishing Company, Chicago.

Waterhouse, R. 1995, *Private Pleasures, Public Leisure: A History of Australian Popular Culture since 1788*, Longman, Melbourne.

Webb, D. 1994, *Greenberg's Guide to Marbles*, 2nd edition, Greenberg Books, Waukesha, Wisconsin.

Webster, D. 1971, *Early Canadian Pottery*, McClelland and Stewart Limited, Toronto.

Wells, T. 1998, 'Nail Chronology: The Use of Technologically Derived Features', *Historical Archaeology* 32 (2): 78-99.

White, G. 1966, *European and American Dolls and their marks and patents*, Chancellor Press, London.

White, M. E. 1994, *After the Greening: The Browning of Australia*, Kangaroo Press, Kenthurst, NSW.

Williams, L. B. 1977, 'Timber Production in the Otway Region', *Proceedings of the Royal Society of Victoria*, vol. 89 (part 1): 89-98.

Williams, M. 1997, 'Ecology, imperialism and deforestation', in *Ecology and Empire. Environmental History of Settler Societies*, eds. T. Griffiths and L. Robin, Melbourne University Press, Melbourne, pp.169-184.

Williams, R. H. 1982, *Dream Worlds. Mass Consumption in late Nineteenth-Century France*, University of California Press, Berkeley.

Willis, E. 1989, *Medical Dominance. The division of labour in Australian health care*, Allen & Unwin, Sydney.

Wilson, R. L. 1981, *Bottles on the Western Frontier*, The University of Arizona Press, Tucson.

Wolf, E. R. 1982, *Europe and the People Without History*, University of California Press, Berkeley.

Woodhead, E. I., Sullivan, C. and Gusset, G. 1984, *Lighting Devices in the National Reference Collection, Parks Canada*, National Historic Parks and Sites Branch, Parks Canada, Environment Canada.

Wright, R. 1989, *The Bureaucrats' Domain: Space and the Public Interest in Victoria 1836-84*, Oxford University Press, Melbourne.

Interviews

Biddle, J. 2 November 1998.

Chamberlain, E. and G. 7 June 1999.

Gardner, N. 15 November 1998.

Henry, R. 12 June 1999.

Harding, J. 21 May 1999.

Washington, J. 24 May 1999

www.ingramcontent.com/pod-product-compliance
Lightning Source LLC
La Vergne TN
LVHW071215210926
848482LV00015B/816